HIGH INTEGRITY DIE CASTING PROCESSES

HIGH INTEGRITY DIE CASTING PROCESSES

EDWARD J. VINARCIK

JOHN WILEY & SONS, INC.

This book is printed on acid-free paper. ∞

Copyright © 2003 by John Wiley & Sons, New York. All rights reserved

Published by John Wiley & Sons, Inc., Hoboken, New Jersey
Published simultaneously in Canada

No part of this publication may be reproduced, stored in a retrieval system or transmitted in any form or by any means, electronic, mechanical, photocopying, recording, scanning or otherwise, except as permitted under Section 107 or 108 of the 1976 United States Copyright Act, without either the prior written permission of the Publisher, or authorization through payment of the appropriate per-copy fee to the Copyright Clearance Center, Inc., 222 Rosewood Drive, Danvers, MA 01923, (978) 750-8400, fax (978) 750-4470, or on the web at www.copyright.com. Requests to the Publisher for permission should be addressed to the Permissions Department, John Wiley & Sons, Inc., 111 River Street, Hoboken, NJ 07030, (201) 748-6011, fax (201) 748-6008, e-mail: permcoordinator@wiley.com.

Limit of Liability/Disclaimer of Warranty: While the publisher and author have used their best efforts in preparing this book, they make no representations or warranties with respect to the accuracy or completeness of the contents of this book and specifically disclaim any implied warranties of merchantability or fitness for a particular purpose. No warranty may be created or extended by sales representatives or written sales materials. The advice and strategies contained herein may not be suitable for your situation. You should consult with a professional where appropriate. Neither the publisher nor author shall be liable for any loss of profit or any other commercial damages, including but not limited to special, incidental, consequential, or other damages.

For general information on our other products and services or for technical support, please contact our Customer Care Department within the United States at (800) 762-2974, outside the United States at (317) 572-3993 or fax (317) 572-4002.

Wiley also publishes its books in a variety of electronic formats. Some content that appears in print may not be available in electronic books.

Library of Congress Cataloging-in-Publication Data

Vinarcik, Edward J.
   High integrity die casting processes / Edward J. Vinarcik.
     p. cm.
   Includes index.
   ISBN 0-471-20131-6
   1. Die-casting. I. Title.
TS239.V56 2002
671.2′53—dc21

2002009957

Printed in the United States of America

10 9 8 7 6 5 4 3 2 1

To My Dearest Friend

Ad Majorem Dei Gloriam

CONTENTS

Preface	xv
Figures and Tables	xvii

INTRODUCTION 1

1 Introduction to High Integrity Die Casting Processes 3
    1.1 Origins of High Pressure Die Casting 3
    1.2 Conventional High Pressure Die Casting 5
    1.3 Problems with Conventional Die Casting 7
    1.4 Strategies to Improve Die Casting Capabilities 10
    1.5 High Integrity Die Casting Processes 10
          References 11

2 Molten Metal Flow in High Integrity Die Casting Processes ... 13
    2.1 Introduction 13
    2.2 Flow within a Fluid 13
    2.3 Flow at the Metal Fill Front 15
    2.4 Metal Flow in Vacuum Die Casting 19
    2.5 Metal Flow in Squeeze Casting 21
    2.6 Metal Flow in Semi-Solid Metalworking 22

	2.7	Predicting Metal Flow in High Integrity Die Casting Processes	24
		References	24

HIGH INTEGRITY DIE CASTING PROCESSES   27

3		**Vacuum Die Casting**	29
	3.1	Vacuum Die Casting Defined	29
	3.2	Managing Gases in the Die	29
	3.3	Managing Shrinkage in the Die	34
	3.4	Elements of Vacuum Die Casting Manufacturing Equipment	35
	3.5	Applying Vacuum Die Casting	40
		References	42
		Case Studies: Vacuum Die Casting	42
		Introduction	42
		B Post	43
		Transmission Cover	44
		Engine Component Mounting Bracket	45
		Marine Engine Lower Mounting Bracket	46
		Reference	49
4		**Squeeze Casting**	51
	4.1	Squeeze Casting Defined	51
	4.2	Managing Gases in the Die	53
	4.3	Managing Shrinkage in the Die	54
	4.4	Elements of Squeeze Casting Manufacturing Equipment	56
	4.5	Applying Squeeze Casting	57
		References	58
		Case Studies: Squeeze Casting	58

		Introduction	58
		Steering Knuckle	60
		Valve Housing	61
		Steering Column Housing	62
		High Performance Engine Block	63
		References	65
5	**Semi-Solid Metalworking**		**67**
	5.1	Semi-Solid Metalworking Defined	67
	5.2	Managing Gases in the Die	70
	5.3	Managing Shrinkage in the Die	70
	5.4	Microstructures in Semi-Solid Metalworking	71
	5.5	Semi-Solid Metalworking Equipment	72
		5.5.1 Billet-Type Indirect Semi-Solid Metalworking	73
		5.5.2 Thixomolding® Direct Semi-Solid Metalworking	79
	5.6	Applying Semi-Solid Metalworking	82
		References	83
		Case Studies: Aluminum Semi-Solid Metalworking	84
		Introduction	84
		Fuel Rails	84
		Control Arm	88
		Swivel Bracket	89
		Idler Housing	90
		References	91
		Case Studies: Magnesium Semi-Solid Metalworking	91
		Introduction	91
		Automotive Seat Frame	93
		Wireless Telephone Face Plates	95

		Video Projector Case	96
		Camera Housing	97
		Laptop Computer Case	97
		Power Hand Tool Housing	98
		References	100
6	**Thermal Balancing and Powder Die Lubricant Processes**		**101**
	6.1	Thermal Cycling Inherent to High Integrity Die Casting Processes	101
	6.2	Heat Checking and Soldering	102
	6.3	Containing the Effects of Heat Checking and Soldering	103
	6.4	Repercussion of Heat Checking and Soldering Containment Actions	105
	6.5	Thermal Management of High Integrity Die Casting Process Tooling	105
	6.6	Minimization of Thermal Cycling Effects with Powder Lubricants	106
	6.7	Applying Thermal Management Methods in Real World Applications	108
		References	109

DESIGN CONSIDERATIONS FOR HIGH INTEGRITY DIE CASTINGS — 111

7	**Design for Manufacturability of High Integrity Die Castings**		**113**
	7.1	Introduction to Design for Manufacturability	113
	7.2	High Integrity Die Casting Design for Manufacturability Guidelines	113
	7.3	Automotive Fuel Rail Case Study Review	114

		7.3.1	Fuel Rail Functional Requirements	115
		7.3.2	Case Study Analysis Method	115
		7.3.3	Review of the Z-1 Fuel Rail Design	116
		7.3.4	Review of the Z-2 Fuel Rail Design	118
		7.3.5	Further Design for Manufacturability Improvements	121
	7.4	Conclusions of the Case Study		122
		References		123

8 Component Integration Using High Integrity Die Casting Processes — 125

	8.1	Introduction to Component Integration	125
	8.2	Hidden Costs in Every Component	125
	8.3	Analyzing Integration Potential	127
	8.4	Component Integration Using High Integrity Die Casting Processes	127
	8.5	Component Integration Case Study	129
		References	130

9 Value Added Simulations of High Integrity Die Casting Processes — 131

			Introduction to Applied Computer Simulations	131
	9.1			
	9.2		Computer Simulations of High Integrity Die Casting Processes	134
	9.3		Applying Simulations Effectively	136
		9.3.1	Resources	138
		9.3.2	Planning	139
		9.3.3	Coupling Product and Process Simulations	140
	9.4		Commitment	140
	9.5		A Case for Sharing Simulation Data across Organizations	140
			References	141

CONTROLLING QUALITY IN HIGH INTEGRITY DIE CASTING PROCESSES — 143

10 Applying Statistical Process Control to High Integrity Die Casting Processes — 145

- 10.1 Introduction to Statistical Process Control — 145
- 10.2 SPC Characteristic Types — 148
- 10.3 SPC Applied to Dynamic Process Characteristics — 149
- 10.4 Die Surface Temperature Case Study — 151
- 10.5 Applying SPC to High Integrity Die Casting Processes — 154
- References — 155

11 Defects in High Pressure Casting Processes — 157

- 11.1 Introduction — 157
- 11.2 Conventional Die Casting Defects — 157
  - 11.2.1 Surface Defects — 158
  - 11.2.2 Internal Defects — 159
  - 11.2.3 Dimensional Defects — 161
- 11.3 Defects Occurring during Secondary Processing — 161
- 11.4 Defects Unique to Squeeze Casting and Semi-Solid Metalworking — 162
  - 11.4.1 Contaminant Veins — 163
  - 11.4.2 Phase Separation — 165
- 11.5 Predicting Defects — 167
- References — 168

VISIONS OF THE FUTURE — 169

12 Future Developments in High Integrity Die Casting — 171

- 12.1 Continual Development — 171
- 12.2 New High Integrity Die Casting Process Variants — 171
- 12.3 Refinements of Magnesium Alloys — 172

12.4	Emerging Alloys for Use with High Integrity Die Casting Processes	173
12.5	Metal Matrix Composites for Use with High Integrity Die Casting Processes	173
12.6	Reducing Tooling Lead Times	175
12.7	Lost-Core Technologies	176
12.8	Controlled Porosity	177
12.9	Innovations Continue	178
	References	178

STUDY QUESTIONS 181

Appendix A	Common Nomenclature Related to High Integrity Die Casting Processes	201
Appendix B	Recommended Reading	207
	B.1 Books	207
	B.2 Papers	208
	B.3 Periodicals	209
Appendix C	Material Properties of Aluminum	211
	References	211
Appendix D	Die Cast Magnesium Material Properties	215
	Reference	218
Index		219

PREFACE

This book has grown largely out of lectures given for a continuing education seminar titled "Advanced Die Casting Processes" presented at the University of Wisconsin-Milwaukee. It is intended for use as a supplement to such a course and as a reference to practicing process engineers, product engineers, and component designers. The content of this book focuses on presenting the concepts behind advanced die casting technologies, specifically vacuum die casting, squeeze casting, and semi-solid metalworking. Moreover, several sections within the book are dedicated to examining case studies that illustrate the practical nature of these processes.

The book is divided into five distinct sections. The initial two chapters of the book are intended to present the basic concepts related to die casting processes and the flow of molten metal. The second portion of the book examines each of the high integrity casting process along with case studies. Three chapters are dedicated to product design as applied to high integrity die casting processes as well as two chapters focusing on quality and defects. An understanding of the defects and their causes can aid in their avoidance. The final chapter of the book deals with future advances under development.

Included with this book is a compact disk containing Microsoft PowerPoint presentations for each chapter. These presentations can be used for training and teaching purposes, or select slides can be extracted from the presentation for use in engineering proposals, customer education seminars, or marketing presentations. To assist the presenter, detailed speaker notes are available for each chapter slide. To view the chapter notes for any given slide, one must right click the computer mouse and select "speaker notes" while running the slide show.

The author wishes to express his gratitude to several colleagues who provided figures, supporting data, and encouragement that made the publication of this book possible. Specifically, the author wishes to thank Joseph Benedyk, Henry Bakemeyer of Die Casting Design and Consulting, John Jorstad of Formcast, Robert Tracy of Foundarex Corporation, Paul Mikkola of Hitchner Manufacturing, Robert Wolfe of Madison-Kipp Corporation, Craig Nelson of IdraPrince, Rath DasGupta of SPX Contech, Charles Van Schilt of Thixocast, Steve LeBeau of Thixomat, Michael Lessiter of the American Foundry Society, and Matsuru Adachi of Ube Machinery.

FIGURES AND TABLES

Figure 1.1 Diagrams filed with Doehler's patent for a production die casting machine. 4

Figure 1.2 Graphical illustration of a hot-chamber die casting machine. 6

Figure 1.3 Graphical illustration of a cold-chamber die casting machine. 7

Figure 1.4 Casting cycle for cold-chamber die casting. 8

Figure 2.1 Comparison of (*a*) laminar flow and (*b*) turbulent flow. 14

Figure 2.2 Illustration of the experiment demonstrating the difference between (*a*) laminar flow and (*b*) turbulent flow. 15

Figure 2.3 Graphical illustration of planar flow. 16

Figure 2.4 Graphical illustration showing the progression of a die cavity filling with a planar metal front. 17

Figure 2.5 Graphical illustration showing nonplanar flow. 17

Figure 2.6 Graphical illustration showing the progression of a die cavity filling with a nonplanar metal front. 18

Figure 2.7 Graphical illustration showing the progression of nonplanar fill. 18

Figure 2.8 Illustration showing atomized flow typical in conventional die casting. 19

Figure 2.9 Graphical illustration of die fill with atomized metal flow in conventional die casting. 20

xviii   FIGURES AND TABLES

Figure 2.10   Short shots of identical castings illustrating the difference between (*a*) planar filling and (*b*) nonplanar filling.   21
Figure 3.1   Graphical illustration showing the progression of a die cavity filling with (*a*) improper vacuum valve placement and (*b*) proper vacuum valve placement.   33
Figure 3.2   Graphical progression showing liquid metal wave cresting and gas entrapment in the shot sleeve: (*a*) pour hole open; (*b*) pour hole closed; (*c*) wave cresting; (*d*) gases trapped.   34
Figure 3.3   Example operating curve for a vacuum pump.   35
Figure 3.4   Illustration of a rotary vane vacuum pump.   36
Figure 3.5   Examples of portable vacuum systems for use in vacuum die casting.   37
Figure 3.6   Schematic of a corrugated chill-block-type vacuum shut-off valve.   38
Figure 3.7   Experimental test data showing the pressure lag when using a corrugated chill-block shut-off valve (1200 cm^3 volume with 0.4 cm^2 X section).   39
Figure 3.8   Example of a mechanical vacuum shut-off valve.   40
Figure 3.9   Example of a hydraulic vacuum shut-off valve.   41
Figure 3.10   Experimental test data showing the pressure response in using a dynamic vacuum shut-off valve (1200 cm^3 volume with 1.6 cm^2 valve X section).   41
Figure 3.11   Components manufactured using vacuum die casting.   43
Figure 3.12   Automotive B post manufactured using vacuum die casting.   44
Figure 3.13   Transmission cover manufactured using vacuum die casting.   45
Figure 3.14   Vacuum die cast engine component mounting bracket.   46
Figure 3.15   Marine engine using a vacuum die cast lower motor mounting bracket.   47

Figure 3.16 Impact properties obtained from test samples; process comparison: $\frac{1}{4}$-in. impact samples machined from casting. 48

Figure 3.17 Actual impact properties obtained for the vacuum die cast marine engine lower motor mounting brackets: heat treatment benefit. 48

Figure 3.18 Marine engine lower motor mounting bracket manufactured using vacuum die casting. 49

Figure 4.1 Schematic of the squeeze forming process. 52

Figure 4.2 Comparisons of casting pressures to gate velocities for numerous die casting processes. 53

Figure 4.3 Microstructural comparisons between conventional die casting and squeeze casting. 54

Figure 4.4 Graphical illustration showing the progression of a die cavity filling with (a) atomized filling and (b) a planar metal front. 55

Figure 4.5 Comparisons of conventional die casting and squeeze casting. 57

Figure 4.6 Components manufactured using the squeeze casting process. 59

Figure 4.7 Steering knuckle manufactured using the squeeze casting process. 60

Figure 4.8 Squeeze cast valve housing. 62

Figure 4.9 Steering column housing produced using the squeeze casting process. 63

Figure 4.10 Porsche Boxter engine block produced using the squeeze casting process. 64

Table 5.1 Freezing ranges for common die cast aluminum alloys. 68

Figure 5.1 Aluminum billet in the semi-solid state. 68

Figure 5.2 Comparisons of casting pressures and gate velocities for numerous die casting processes. 69

Figure 5.3 Process comparison between (a) direct semi-solid metalworking, (b) indirect semi-solid metalworking, and (c) conventional casting processes. 72

Figure 5.4	Microstructure of an aluminum component produced with a direct semi-solid metalworking process.	73
Figure 5.5	Microstructure of an aluminum component produced with an indirect semi-solid metalworking process.	74
Figure 5.6	Graphical representation of a typical indirect semi-solid metalworking manufacturing cell.	75
Figure 5.7	Continuously cast semi-solid metalworking feedstock.	76
Figure 5.8	Anatomy of a continuously cast semi-solid metalworking billet (*a*) before heating and (*b*) after heating.	77
Figure 5.9	Plunger tip and die design for capturing the dendritic case of a continuously cast semi-solid metalworking billet.	78
Figure 5.10	Anatomy of an extruded semi-solid metalworking billet (*a*) before heating and (*b*) after heating.	78
Figure 5.11	Schematic of Thixomolding® machine use in direct semi-solid metalworking.	79
Figure 5.12	Graphical representation of a typical manufacturing cell.	80
Figure 5.13	Schematic of the metal injection screw used in Thixomolding®.	81
Figure 5.14	Die design for capturing the frozen screw plug.	81
Figure 5.15	Multiple microstructures may be obtained by varying the percent solid during metal injection when using the Thixomolding® process.	82
Figure 5.16	Comparisons between conventional die casting, semi-solid metalworking, and squeeze casting.	83
Figure 5.17	Automotive components manufactured using semi-solid metalworking processes.	85
Figure 5.18	Components manufactured using semi-solid metalworking processes.	85
Figure 5.19	Fuel rails manufactured using semi-solid metalworking (bottom) and brazing (top).	86

FIGURES AND TABLES xxi

Figure 5.20   Fuel rail for use on 2.0-liter and 2.2-liter engines manufactured using an indirect semi-solid metalworking process.   87
Figure 5.21   Control arm manufactured using an indirect semi-solid metalworking processes.   88
Figure 5.22   Outboard motor swivel bracket manufactured using semi-solid metalworking.   89
Figure 5.23   Automotive idler housing manufactured using an indirect semi-solid metalworking process.   90
Figure 5.24   Magnesium die casting shipments in the United States, 1990–2001.   92
Figure 5.25   Magnesium components manufactured using semi-solid metalworking processes.   93
Figure 5.26   Magnesium automotive seat frame manufactured using semi-solid metalworking.   94
Figure 5.27   Wireless telephone face plates manufactured in magnesium using semi-solid metalworking.   95
Figure 5.28   Magnesium LCD projector case manufactured in three sections using semi-solid metalworking.   96
Figure 5.29   Hand-held video camera housing produced using a magnesium semi-solid metalworking process.   98
Figure 5.30   Compact laptop personal computer case manufactured using semi-solid metalworking in lightweight magnesium.   99
Figure 5.31   Power hand tool housing produced in magnesium using semi-solid metalworking.   100
Figure 6.1   Die cavity surface temperature over a single processing cycle.   102
Figure 6.2   Die surface cross section illustrating heat-checking formation and progression.   103
Figure 6.3   Die surface cross section illustrating microstructural weaknesses in a welded die surface.   104
Figure 6.4   Shot sleeve and tool design for use with closed die powder lubricant application.   107
Figure 6.5   Processing cycle for closed die powder lubricant application.   108

xxii   FIGURES AND TABLES

Figure 7.1	Production fuel rails Z-1 (top) and Z-2 (bottom).	114
Figure 7.2	Diagram of Z-1 fuel rail design.	116
Figure 7.3	Shrinkage porosity found in the regulator pocket of the Z-1 fuel rail.	117
Figure 7.4	Diagram of Z-2 fuel rail design.	118
Figure 7.5	Z-2 end-flange evolution: (*a*) initial asymmetric design; (*b*) symmetric design; (*c*) symmetric webbed design.	120
Figure 7.6	Shrinkage porosity and contaminant vein oxide inclusions found in the second flange design of Z-2 fuel rail.	120
Figure 7.7	Shrinkage porosity found in the body of a production Z-2 fuel rail.	122
Figure 8.1	Component integration analysis flow chart.	128
Figure 8.2	Four-cylinder fuel rail produced by brazing a fabricated assembly.	129
Figure 8.3	Four-cylinder fuel rail produced using the semi-solid metalworking process.	130
Figure 9.1	Control factors and potential problems in the product development cycle.	132
Figure 9.2	Project cost lever illustrating returns as a function of when an investment is made.	133
Figure 9.3	Computer simulation of die filling during metal injection.	135
Figure 9.4	Computer simulation illustrating areas in the die cavity prone to solidification shrinkage porosity.	136
Figure 9.5	Computer simulation showing variation in residual stress that forms during solidification and cooling.	137
Figure 9.6	Computer simulation showing component distortion (exaggerated) during cooling.	138
Figure 9.7	Qualitative illustration grouping the number of organizations to the effectiveness of their simulation efforts.	139
Figure 10.1	Dart board comparison of (*a*) common-cause variation and (*b*) special-cause variation.	146
Figure 10.2	Dart board comparison showing a reduction in common-cause variation from (*a*) to (*b*).	147

Figure 10.3   Example $\bar{x}$ chart commonly used for SPC.   149
Figure 10.4   Process data curve for die surface temperature over one cycle.   151
Figure 10.5   Thirty discrete process data curves for die surface temperature over one cycle (equal number of elements for each cycle).   152
Table 10.1   Calculation of upper and lower control limits for element 0.0.   153
Figure 10.6   Process average curve, upper control limit curve, and lower control limit curve for die surface temperature over one cycle.   154
Figure 10.7   Process average curve, upper control limit curve, and lower control limit curve for die surface temperature during die lubricant spray.   154
Figure 10.8   Process data curve exhibiting an "out-of-control" condition (cycle exhibits special-cause variation).   155
Figure 11.1   Contaminant veins form as (a) a clean planar fill front collects contaminants and (b) the metal progresses through the die cavity.   163
Figure 11.2   Characteristic features prone to contaminant vein formation.   164
Figure 11.3   Graphical illustration showing the effects of phase separation in semi-solid metalworking.   166
Figure 12.1   Four-step process for producing a semi-solid slurry on demand.   172
Figure 12.2   Motor cycle sprocket die cast using an SiC particulate reinforced aluminum matrix composite.   174
Figure 12.3   Hollow (a) aluminum automotive suspension arm and (b) resin-bonded sand core.   177
Figure SQ1   Three different types of metal flow behavior.   181
Figure S1   Illustration of (a) atomized metal flow, (b) nonplanar metal flow, and (c) planar metal flow.   182
Figure SQ2   Illustration of metal behavior in three processes requiring vacuum valve placement.   183
Figure S2   Illustrations showing optimum vacuum valve placement for three metal flow patterns.   184

xxiv FIGURES AND TABLES

Figure SQ3   Graphical representation of a typical indirect semi-solid metalworking manufacturing cell.   185
Figure S3   Processing order numbered for a typical indirect semi-solid metalworking manufacturing cell.   186
Figure SQ4   Graphical representation of a typical Thixomolding® manufacturing cell.   187
Figure S4   Processsing order numbered for a typical Thixomolding® manufacturing cell.   188
Figure SQ5   Anatomy of (*a*) continuously cast and (*b*) extruded semi-solid metalworking billet.   189
Figure S5   (*a*) Continuously cast and (*b*) extruded semi-solid metalworking billet behavior during heating.   190
Figure SQ6   Four aluminum microstructures produced using different die casting processes.   191
Figure S6   Four aluminum microstructures with die casting process method noted.   192
Figure SQ7   Die surface temperature over one casting cycle.   193
Figure S7   Die surface temperature over one casting cycle with casting phases noted.   194
Figure SQ8   Microstructural regions of a welded die face.   195
Figure S8   Microstructural make-up of a welded die face.   196
Figure SQ9   Air scoop for an agriculture combine produced as a welded facrication.   197
Figure S9.1 Illustration noting the seven individual components of the fabricated air scoop.   198
Figure S9.2 Illustration noting 24 weld points on the fabricated air scoop.   198
Figure S9.3 Integrated single piece air scoop.   199
Table C.1   Tensile properties of common die cast aluminum alloys.   212
Table C.2   Impact properties of common die cast aluminum alloys.   212
Table C.3   Wear and cavitation resistance of common die cast aluminum alloys.   213

Table C.4	Impact resistance and properties of common die cast aluminum alloys.	213
Table C.5	Fracture toughness of common die cast aluminum alloys.	214
Table C.6	Fatigue properties of common die cast aluminum alloys.	214
Table D.1	Material properties for magnesium alloys.	216

HIGH INTEGRITY DIE CASTING PROCESSES

INTRODUCTION

1
INTRODUCTION TO HIGH INTEGRITY DIE CASTING PROCESSES

1.1 ORIGINS OF HIGH PRESSURE DIE CASTING

Casting processes are among the oldest methods for manufacturing metal goods. In most early casting processes (many of which are still used today), the mold or form used must be destroyed in order to remove the product after solidification. The need for a permanent mold, which could be used to produce components in endless quantities, was the obvious alternative.

In the Middle Ages, craftsmen perfected the use of iron molds in the manufacture of pewterware. Moreover, the first information revolution occurred when Johannes Gutenberg developed a method to manufacture movable type in mass quantities using a permanent metal mold. Over the centuries, the permanent metal mold processes continued to evolve. In the late 19th century processes were developed in which metal was injected into metal dies under pressure to manufacture print type. These developments culminated in the creation of the linotype machine by Ottmar Mergenthaler. However, the use of these casting methods could be applied to manufacture more than type for the printing press.

H. H. Doehler is credited with developing die casting for the production of metal components in high volumes. Shown in Figure 1.1 are diagrams filed with patent 973,483 for his first production die casting machine.[1] Initially, only zinc alloys were used in die casting. Demands for other metals drove the development

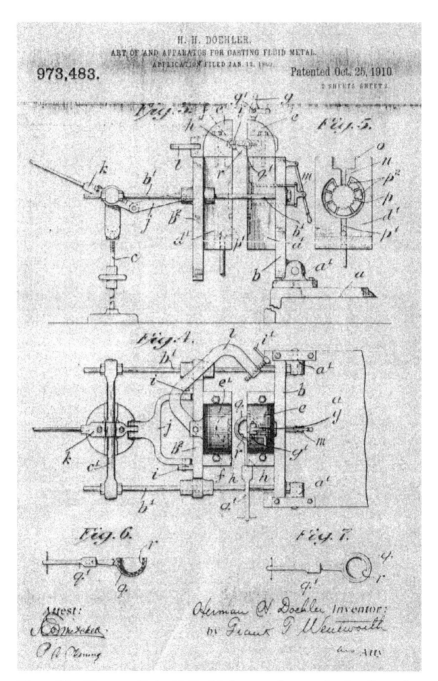

Figure 1.1 Diagrams filed with Doehler's patent for a production die casting machine.

of new die materials and process variants. By 1915, aluminum alloys were being die cast in large quantities.[2]

Much progress has been made in the development of die casting technologies over the last century. Developments continue to be made driving the capabilities of the process to new levels and increasing the integrity of die cast components.

1.2 CONVENTIONAL HIGH PRESSURE DIE CASTING

Conventional die casting (CDC) is a net-shape manufacturing process using a permanent metal die that produces components ranging in weight from a few ounces to nearly 25 kg quickly and economically. Traditionally, die casting is not used to produce large products; past studies, however, have shown that very large products, such as a car door frame or transmission housing, can be produced using die casting technologies.[2] Conventional die cast components can be produced in a wide range of alloy systems, including aluminum, zinc, magnesium, lead, and brass.

Two basic conventional die casting processes exist: the hot-chamber process and the cold-chamber process. These descriptions stem from the design of the metal injection systems utilized.

A schematic of a hot-chamber die casting machine is shown in Figure 1.2. A significant portion of the metal injection system is immersed in the molten metal at all times. This helps keep cycle times to a minimum, as molten metal needs to travel only a very short distance for each cycle. Hot-chamber machines are rapid in operation with cycle times varying from less than 1 sec for small components weighing less than a few grams to 30 sec for castings of several kilograms. Dies are normally filled between 5 and 40 msec. Hot-chamber die casting is traditionally used for low melting point metals, such as lead or zinc alloys. Higher melting point metals, including aluminum alloys, cause rapid degradation of the metal injection system.

Cold-chamber die casting machines are typically used to conventionally die cast components using brass and aluminum alloys. An illustration of a cold-chamber die casting machine is presented in Figure 1.3. Unlike the hot-chamber machine, the metal injection system is only in contact with the molten metal for a short period

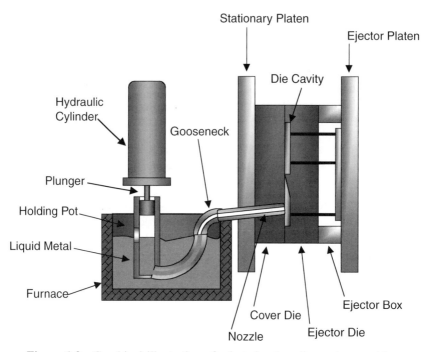

Figure 1.2 Graphical illustration of a hot-chamber die casting machine.

of time. Liquid metal is ladled (or metered by some other method) into the shot sleeve for each cycle. To provide further protection, the die cavity and plunger tip normally are sprayed with an oil or lubricant. This increases die material life and reduces the adhesion of the solidified component.

All die casting processes follow a similar production cycle. Figure 1.4 is an illustration of the cycle using the cold-chamber die casting process as a model. Initially, liquid metal is metered into an injection system (*a*), which is then immediately pushed (*b*) through a runner system (*c*) into a die cavity (*d*) under high pressure. High pressures are maintained on the alloy during solidification. After complete solidification, the die opens (*e*) and the component is ejected (*f*).

Conventional die casting is an efficient and economical process. When used to its maximum potential, a die cast component may replace an assembly composed of a variety of parts produced by various manufacturing processes. Consolidation into a single die casting can significantly reduce cost and labor.

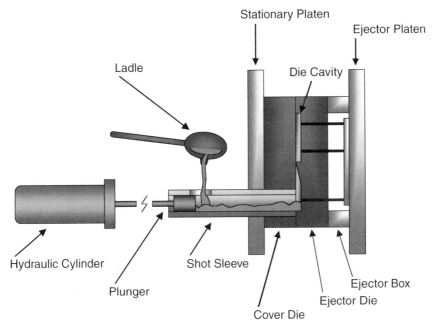

Figure 1.3 Graphical illustration of a cold-chamber die casting machine.

1.3 PROBLEMS WITH CONVENTIONAL DIE CASTING

Conventional die casting is utilized to produce many products in the current global market. Unfortunately, conventional die casting has a major limitation that is preventing its use on a broader scale. A potential defect, commonly found in conventionally die cast components, is porosity.

Porosity often limits the use of the conventional die casting process in favor of products fabricated by other means. Pressure vessels must be leak tight. Conventional die castings often are unable to meet this requirement. Moreover, the detection of porosity is difficult. In some cases, an "as-produced" component is acceptable. Subsequent machining, however, cuts into porosity hidden within the component, compromising the integrity of the product.

Porosity is attributed to two main sources: solidification shrinkage and gas entrapment. Most alloys have a higher density in their solid state as compared to their density in the liquid state. As a result, shrinkage porosity forms during solidification. Due to the

8 INTRODUCTION TO HIGH INTEGRITY DIE CASTING PROCESSES

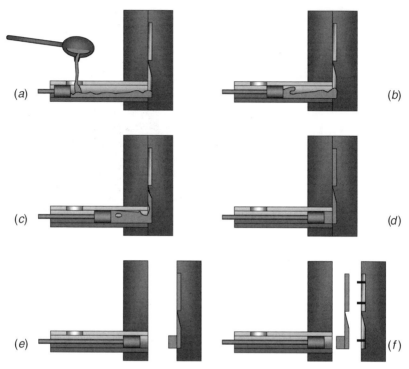

Figure 1.4 Casting cycle for cold-chamber die casting.

turbulent manner in which metal enters and fills the die cavity, gas often becomes entrapped in the metal, resulting in porosity.

Porosity also affects the mechanical properties of conventionally die cast components. In structural applications, porosity can act as a stress concentrator creating an initiation site for cracks.

Numerous studies have documented how porosity in die castings varies with several operating conditions.[3–8] A method has been developed for quantifying the porosity in die cast components.[9] The total porosity contained in a component is defined using the equation

$$\%P = \text{(solidification shrinkage)} + \text{(gas contribution)} \quad (1.1)$$

which can be further defined as

$$\%P = \frac{\beta V^*}{V_c} + \left(\phi \frac{T\rho L}{(237\ K)P}\right)(\nu - \nu^*) \qquad (1.2)$$

where

$\%P$ = percent porosity,
β = solidification shrinkage factor in percent,
V^* = volume of liquid in casting cavity that is not supplied liquid during solidification in cubic centimeters,
V_c = volume of the casting cavity in cubic centimeters,
T = temperature of the gas in the casting cavity in degrees Kelvin,
P = pressure applied to the gas during solidification in atmospheres,
ϕ = fraction of the gas that does not report to the solidification shrinkage pores,
ρ = liquid alloy density at the melting temperature in grams per cubic centimeter,
ν = quantity of the gas contained in the casting at standard temperature and pressure conditions (273 K at 1 atm) in cubic centimeters per 100 g of alloy, and
ν^* = solubility limit of gas in the solid at the solidus temperature at standard temperature and pressure conditions in cubic centimeters per 100 g of alloy.

The first portion of Equation 1.2 is a relationship for porosity due to solidification shrinkage. The second portion of Equation 1.2 describes the porosity due to gas entrapment. The total gas contained in the casting includes gas from physical entrapment, gas from lubricant decomposition, and gas dissolved in the alloy. This relationship can also be described mathematically,

$$\nu = \nu_{\text{Entrained}} + \nu_{\text{Lube}} + \nu_{\text{Soluble gas}} \qquad (1.3)$$

Each of the gas contributions in Equation 1.3 is expressed in cubic centimeters at standard temperature and pressure conditions per 100 g of alloy.

In addition to porosity, the microstructures inherent with the conventional die casting cannot meet the mechanical requirements needed for many applications. Subsequent heat treating, which can alter the microstructure, is rarely possible due to defects that emerge during thermal processing, such as blistering.

Regardless of the limitations found in conventional die cast components, demands exist for high integrity products. In many cases, product engineers and designers turn to investment casting, forging, injection molding, and assembled fabrications to meet necessary requirements. Typically, these processes are more costly than conventional die casting in both processing time and raw material costs.

1.4 STRATEGIES TO IMPROVE DIE CASTING CAPABILITIES

Several efforts have proven successful in stretching the capabilities of conventional die casting while preserving short cycle times and providing dimensional stability and other beneficial characteristics. In these efforts, three strategies have extended the capabilities of the die casting process:

1. eliminating or reducing the amount of entrapped gases,
2. eliminating or reducing the amount of solidification shrinkage, and
3. altering the microstructure of the metal.

The first two strategies noted affect each of the major quantities that contribute to porosity as defined in Equation 1.1. The third strategy addresses the mechanical properties by modifying the fundamental structure of the die cast component.

1.5 HIGH INTEGRITY DIE CASTING PROCESSES

Three high integrity die casting processes have been successfully developed and deployed for commercial use in high volume production. These processes are vacuum die casting, squeeze casting, and semi-solid metalworking (SSM).

Vacuum die casting utilizes a controlled vacuum to extract gases from the die cavities and runner system during metal injection. This process works to minimize the quantities of $\nu_{\text{Entrained}}$ and ν_{Lube} as defined in Equation 1.3. Porosity due to entrapped gases is virtually eliminated.

Squeeze casting is characterized by the use of a large gate area and planar filling of the metal front within the die cavity. As with vacuum die casting, this process works to minimize the quantities of $\nu_{\text{Entrained}}$ and ν_{Lube} as noted in Equation 1.3. The mechanism, however, is much different. Planar filling allows gases to escape from the die, as vents remain open throughout metal injection. Furthermore, the large gate area allows metal intensification pressure to be maintained throughout solidification, reducing the magnitude of V^* as defined in Equation 1.2. Both porosity from entrapped gas and solidification shrinkage are reduced by using squeeze casting.

Semi-solid metalworking is the most complex of the high integrity die casting processes. During semi-solid metalworking a partially liquid–partially solid metal mixture is injected into the die cavity. The fill front is planar, minimizing gas entrapment, as in squeeze casting. Moreover, solidification shrinkage is greatly reduced, as a significant portion of the metal injected into the die cavity is already solid. Semi-solid metalworking addresses both sides of the porosity relationship defined in Equation 1.1.

In addition to reducing porosity, a unique microstructure is generated during semi-solid metalworking. The mechanical properties inherent to this microstructure are superior to those created in conventionally die cast components.

Products produced using high integrity die casting processes have little or no porosity. Moreover, the mechanical properties are much improved in comparison to conventional die cast components. This is due to reduced levels of porosity, the viability of subsequent heat treating, and formation of microstructures not possible with the conventional die casting process.

REFERENCES

1. Doehler, H., "Art of and Apparatus for Casting Fluid Metal," United States Patent 973,483, United States Patent and Trademark Office, Washington, D.C., 25, October 1910.

2. Doehler, H., *Die Casting,* McGraw Hill Book Company, New York, 1951.
3. Lindsey, D., and Wallace, J., "Effect of Vent Size and Design, Lubrication Practice, Metal Degassing, Die Texturing and Filling of Shot Sleeve on Die Casting Soundness," *Proceedings 7th International Die Casting Congress,* 1972, pp. 1–15.
4. Hayes, D., "Plunger Lubricants Are Important Too!" *Die Casting Engineer,* November/December 1983, p. 32.
5. Gordon, A., Meszaros, G., Naizer, J., Gangasani, P., and Mobley, C., *Comparison of Methods for Characterizing Porosity in Die Castings,* Report No. ERC/NSM-91-51-C, The Ohio State University Engineering Research Center for Net Shape Manufacturing, August 1991.
6. Meszaros, G., *Lubricant Gasification as a Contributing Factor to Porosity in Die Casting,* Masters Thesis, The Ohio State University, Columbus, 1992.
7. Gordon, A., *The Effects of Porosity on the Tensile Properties of Die Cast Aluminum Alloys B390 and B380,* Master's Thesis, The Ohio State University, Columbus, 1992.
8. Vinarcik, E., and Mobley, C., *Decomposition and Gasification Characteristics of Die Casting Plunger Lubricants,* Report No. ERC/NSM-UIRS-92-17, The Ohio State University Engineering Research Center for Net Shape Manufacturing, October 1992.
9. Gordon, A., Meszaros, G., Naizer, J., and Mobley, C., *A Method for Predicting Porosity in Die Castings,* Technical Brief No. ERC/NSM-TB-91-04-C, The Ohio State University Engineering Research Center for Net Shape Manufacturing, September 1991.

2
MOLTEN METAL FLOW IN HIGH INTEGRITY DIE CASTING PROCESSES

2.1 INTRODUCTION

Many of the defects and problems encountered when producing components using the conventional die casting process stem from the type of liquid metal flow within the die cavity. Each high integrity die casting process combats these issues differently. In the cases of squeeze casting and semi-solid metalworking, the type of metal flow within the die cavity is completely different than that in conventional and vacuum die casting. Before exploring each of the high integrity die casting processes, a foundation is necessary in the different types of liquid metal flow.

2.2 FLOW WITHIN A FLUID

The nature of fluid flow is critically dependent on the velocity and physical properties of the fluid. At low flow rates, a fluid moves in a stable line parallel to the direction of flow. The flow can be regarded as the unidirectional movement of lamellae with no mixing of the fluid. This phenomenon is known as laminar flow, as shown in Figure 2.1a. For high flow rates, the stability of the fluid is compromised and macroscopic mixing results, as shown in Figure 2.1b. This type of flow is known as turbulent flow. For the

14 MOLTEN METAL FLOW IN HIGH INTEGRITY DIE CASTING PROCESSES

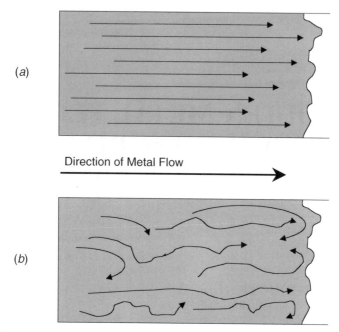

Figure 2.1 Comparison of (*a*) laminar flow and (*b*) turbulent flow.

flow of a fluid with specific properties in a conduit with a specific geometry, a critical velocity exists below which the flow is laminar and above which the flow is turbulent. This phenomenon was ascertained as the result of studies conducted in 1883 by Osborne Reynolds.[1]

Figure 2.2 graphically presents the experiments conducted by Reynolds in which a colored dye was injected into a liquid flowing in a glass tube. At low flow rates, the dye flowed with the fluid without mixing, as shown in Figure 2.2*a*. When increasing the flow of the fluid (Figure 2.2*b*), the colored dye rapidly broke up and mixed with the fluid from its point of injection. As a result of these experiments, Reynolds established the criterion for the transition from laminar to turbulent flow in terms of the dimensionless quantity presented in the following equation:

$$\text{Re} = Dv\rho/\eta \qquad (2.1)$$

in which D is the characteristic geometry of the conduit and v, ρ,

2.3 FLOW AT THE METAL FILL FRONT

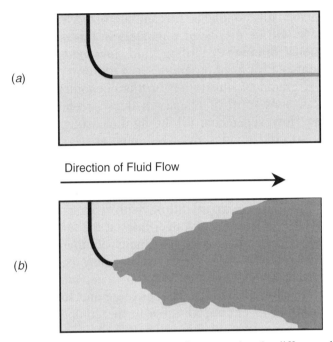

Figure 2.2 Illustration of the experiment demonstrating the difference between (a) laminar flow and (b) turbulent flow.

and η are the velocity, density, and viscosity of the fluid, respectively. This dimensionless number (Re) is known as the Reynolds number.

Turbulent flow is often perceived as detrimental to casting processes while laminar-type flow is preferred. This misconception stems from the confusion between liquid metal flow and the liquid metal fill front that progresses within a mold or die cavity. The liquid metal fill front has a much greater effect on casting integrity than the type of flow within the bulk liquid metal.

2.3 FLOW AT THE METAL FILL FRONT

Although an understanding of the bulk liquid metal flow is often useful, the flow of the metal at the fill front is of most concern. Three distinct metal fill fronts are encountered in die casting processes: planar fill, nonplanar fill, and atomized fill. Each of these phenomena will be discussed separately in this section.

Traditional thinking regarding fluid flow tends to assume that the liquid metal fill front progresses as a uniform plane throughout the die. A graphical illustration of this phenomenon is shown in Figure 2.3. This form of planar fill does occur in some casting processes. However, planar fill during die casting occurs only under very specific conditions. The complex geometries of most components cause the liquid metal fill front to separate. When planar filling of the die cavity does occur, gases trapped within the die are pushed ahead of the metal fill front. In Figure 2.4, the progression of a die cavity filling with a planar metal front is shown. By locating vents and overflows at the farthest point from the gate, gas entrapment can be virtually eliminated.

Nonplanar flow is also observed in many casting processes. Unlike planar metal flow, the fill front is not uniform, as shown in Figure 2.5. Often metal fronts converge and surround a pocket of air, resulting in entrapped gases in the component being produced. When die casting, nonplanar fill often results in the die cavity being filled from the outside inward. This fill behavior is shown in Figures 2.6 and 2.7. In Figure 2.6, the metal front enters the die as a single stream that changes direction only after contacting the far side of the die cavity. The metal front continues to hug the surface of the die, filling the cavity from the outside in creating large pockets of entrapped gas. In Figure 2.7, the metal stream begins to fan out after entering the die cavity. As the metal

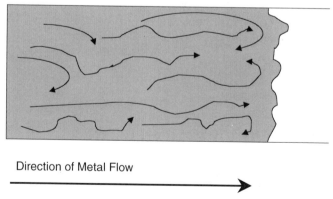

Direction of Metal Flow

Figure 2.3 Graphical illustration of planar flow.

2.3 FLOW AT THE METAL FILL FRONT 17

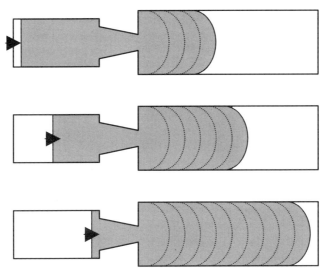

Figure 2.4 Graphical illustration showing the progression of a die cavity filling with a planar metal front.

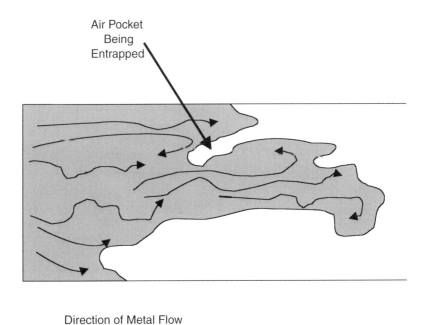

Figure 2.5 Graphical illustration showing nonplanar flow.

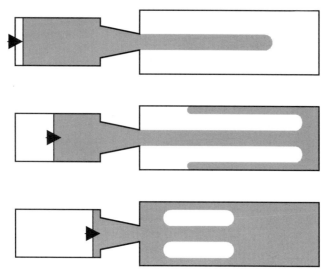

Figure 2.6 Graphical illustration showing the progression of a die cavity filling with a nonplanar metal front.

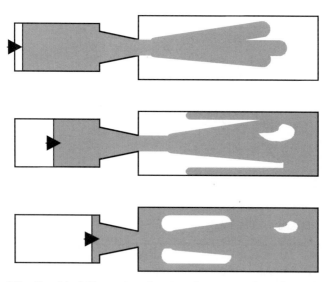

Figure 2.7 Graphical illustration showing the progression of nonplanar fill.

reaches the far side of the die cavity, gases are entrapped as the fill front doubles over on itself. The metal front continues to travel along the surface of the die, filling the cavity from the outside inward. This results in additional pockets of entrapped gas.

When liquid metal is traveling at high velocities through a very small gate, the fill front breaks down, resulting in atomization. This phenomenon is illustrated in Figure 2.8. Due to the high pressure and velocities, the metal becomes in effect an aerosol, spraying into the die cavity. Shown in Figure 2.9 is the progression of die fill, which occurs with atomized metal flow. Liquid metal is sprayed into the die. Filling occurs from the surface of the cavity inward. Typically, the first metal to enter the die strikes the far side of the die cavity and solidifies immediately.

2.4 METAL FLOW IN VACUUM DIE CASTING

In conventional die casting, high gate velocities result in atomized metal flow within the die cavity, as shown in Figures 2.8 and 2.9. Entrapped gas is unavoidable. This phenomenon is also present in vacuum die casting, as the process parameters are virtually identical to that of conventional die casting.

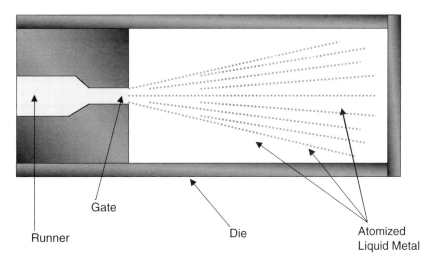

Figure 2.8 Illustration showing atomized flow typical in conventional die casting.

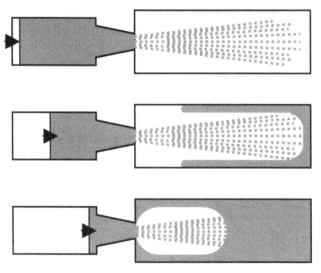

Figure 2.9 Graphical illustration of die fill with atomized metal flow in conventional die casting.

Example Calculation 2.4

Using the vacuum die casting process, a component is manufactured with a conventional aluminum alloy. Calculate the Reynolds number for this process to determine if metal flow at the gate is laminar or turbulent given that the gate is 2 mm wide and 35 mm in length. The velocity of metal at the gate is 50,000 cm/sec. The density and viscosity of liquid aluminum are 2.7 g/cm^3 and 1×10^{-3} g/cm · sec, respectively.[2]

Solution

For this case, the characteristic length is the width of the gate. Using Equation 2.1,

$$\text{Re} = \frac{Dv\rho}{\eta}$$

$$= \frac{(0.2 \text{ cm})(50{,}000 \text{ cm/sec})(2.7 \text{ g/cm}^3)}{1 \times 10^{-3} \text{ g/cm} \cdot \text{sec}}$$

$$= 27{,}000{,}000$$

Due to the extremely high Reynolds number (>10,000), fluid flow at the gate in this vacuum die casting example is extremely turbulent.

2.5 METAL FLOW IN SQUEEZE CASTING

Due to larger gate cross sections and longer fill times in comparison to conventional die casting, atomization of the liquid metal is avoided when squeeze casting. Both planar and nonplanar flows occur in squeeze casting. Achieving planar flow, however, is dependent on the die design and optimization of the process parameters. Figure 2.10 is a picture showing two short shots of identical castings. In Figure 2.10a planar filling occurred within

(a)   (b)

Figure 2.10 Short shots of identical castings illustrating the difference between (a) planar filling and (b) nonplanar filling. (Courtesy of Formcast, Inc.)

the die, while nonplanar filling occurred in Figure 2.10b. These differences in metal flow were made possible by adjusting machine-controlled process parameters. Be that as it may, for complex component geometries, nonplanar fill may be unavoidable.

Example Calculation 2.5

Utilizing a common commercial aluminum alloy, a component is manufactured using squeeze casting technology. Parameters for this process include a gate velocity of 500 cm/sec and circular gate diameter of 10 mm. Determine if the liquid metal flow through the gate is laminar. The density and viscosity of liquid aluminum are 2.7 g/cm^3 and 1×10^{-3} g/cm · sec, respectively.[2]

Solution

Using Equation 2.1, the Reynolds number may be used to determine if the metal flow is laminar at the gate. Since the gate geometry is circular, the characteristic length is the diameter of the gate[3]:

$$Re = \frac{Dv\rho}{\eta}$$

$$= \frac{(1.0 \text{ cm})(500 \text{ cm/sec})(2.7 \text{ g/cm}^3)}{1 \times 10^{-3} \text{ g/cm} \cdot \text{sec}}$$

$$= 1{,}350{,}000$$

For fluid flow through a circular cross section, the transition from laminar to turbulent flow is completed when the Reynolds number reaches 3000.[3] Fluid flow at the gate is not laminar. Flow is turbulent for this squeeze casting example.

2.6 METAL FLOW IN SEMI-SOLID METALWORKING

Semi-solid metalworking is often incorrectly sighted as exhibiting laminar flow when filling the die cavity.[4-6] This misconception has been proliferated in the sales and marketing of semi-solid metalworking related products. Regardless of the increased viscosity of semi-solid metal mixtures, the high flow rates encountered when filling the die cavity under production conditions result in turbulence. Turbulence, however, does not cause gases to be entrapped in the metal. Entrapment of gases occurs at the metal fill front.

Semi-solid metalworking exhibits planar metal flow as the die cavity is filled. This is illustrated in Figures 2.3 and 2.4. Planar filling in semi-solid metalworking processes is the mechanism that reduces the entrapment of gases, not laminar flow. Although planar filling of the die cavity may solve a chronic problem encountered with many casting processes, this phenomenon can lead to unique defects, which will be presented in Chapter 11.

Example Calculation 2.6

When producing an aluminum component using a semi-solid metalworking process, the gate of the die cavity is circular with a diameter of 0.5 cm. Process parameters were optimized with a metal flow velocity of 500 cm/sec at the gate. The density and viscosity of semi-solid aluminum are 2.7 g/cm^3 and 1×10^{-1} g/cm · sec, respectively.[2] Determine if the liquid metal flow through the gate is turbulent or laminar.

Solution

In order to determine if the metal flow is laminar or turbulent, the Reynolds number for the system must be calculated using Equation 2.1. The characteristic length is the diameter of the gate:

$$\text{Re} = \frac{Dv\rho}{\eta}$$

$$= \frac{(0.5 \text{ cm})(500 \text{ cm/sec})(2.7 \text{ g/cm}^3)}{1 \times 10^{-1} \text{ g/cm} \cdot \text{sec}}$$

$$= 6750$$

The turbulent flow is present in fluid flow through a circular crosssection when the Reynolds number exceeded 3000.[3] In this case, the metal flow is turbulent.

2.7 PREDICTING METAL FLOW IN HIGH INTEGRITY DIE CASTING PROCESSES

Over the last decade vast improvements in computer hardware and software technology have made complex simulations of physical phenomena possible. Today engineers and designers have available an ever-growing and continually refined set of tools to aid in product development. Mathematical models using both finite element and finite difference techniques have been developed to simulate various manufacturing processes. Specific to high integrity die casting processes, mathematical models have been developed to simulate several elements, including

mold filling,
air entrapment,
liquid metal surface tracking (for predicting inclusion locations),
solidification thermodynamics,
material properties after solidification,
shrinkage porosity, and
part distortion.

Process modeling can be used to predict and design desired flow conditions within the die cavity. This can yield significant returns on investment by optimizing the manufacturing process prior to building dies.

When utilizing computer models, one must consider that two phases are present in semi-solid metalworking. The currently available computational fluid dynamic software defines the semi-solid metal as a high viscosity fluid rather than as a true two-phase mixture. A method of modeling two-phase flow was proposed in 1997, and efforts are currently underway to develop the proposal into a viable computer model.[7]

REFERENCES

1. Reynolds, O., "An Experimental Investigation of the Circumstances Which Determine Whether Motion of Water Shall Be Direct or Sinuous and of the

REFERENCES

Law of Resistance in Parallel Channels," *Transactions of the Royal Society of London*, Vol. A174, 1883, p. 935.

2. Flemings, M., "Behavior of Metal Alloys in the Semisolid State," *Metallurgical Transactions,* Vol. 22B, June 1991, p. 269.

3. Gaskell, D., *An Introduction to Transport Phenomena in Materials Engineering,* Macmillan, New York, NY, 1992.

4. Keeney, M., J. Courtois, R. Evans, G. Farrior, C. Kyonka, A. Koch, K. Young. "Semisolid Metal Casting and Forging," in Stefanescu, D. (editor), *Metals Handbook,* 9th ed., Vol. 15, *Casting,* ASM International, Materials Park, OH, 1988, p. 327.

5. Young, K., "Semi-solid Metal Cast Automotive Components: New Markets for Die Casting," Paper Cleveland T93-131, North American Die Casting Association, Rosemont, IL, 1993.

6. Siegert, K. and R. Leiber. "Thixoforming of Aluminum," SAE Paper Number 980456, Society of Automotive Engineers, Warrendale, PA, 1998.

7. Alexandrou, A., G. Burgos, and V. Entov "Semisolid Metal Processing: A New Paradigm in Automotive Part Design," SAE Paper Number 2000-01-0676, Society of Automotive Engineers, Warrendale, PA, 2000.

HIGH INTEGRITY DIE CASTING PROCESSES

3
VACUUM DIE CASTING

3.1 VACUUM DIE CASTING DEFINED

Entrapped gas is a major source of porosity in conventional die castings. Vacuum die casting is characterized by the use of a controlled vacuum to extract gases from the die cavities, runner system, and shot sleeve during processing.[1] This high integrity process stretches the capabilities of conventional die casting while preserving its economic benefits.

Numerous metal casting processes have utilized vacuum systems to assist in the removal of unwanted gases. These processes include permanent-mold casting, lost-foam casting, plaster mold casting, and investment casting. The constraint in the evolution of vacuum die casting has been the development of a reliable vacuum shut-off valve. Vacuum die casting is compatible with other high integrity processes, including squeeze casting and semi-solid metalworking.

The integrity of vacuum die cast components is much improved in comparison to conventional die casting. This is due to reduced porosity levels made possible by the minimization of entrapped gas, as discussed in the following section.

3.2 MANAGING GASES IN THE DIE

Gas porosity can originate from many sources, including the physical entrapment of gas during die filling, the decomposition of manufacturing lubricants, and the evolution of gases dissolved in the liquid alloy.[2] The vacuum die casting process minimizes gas entrapment by removing gases from the cavity generated by two of these mechanisms. Both air in the die cavity and gases generated by the decomposition of lubricants can be removed using the vacuum die casting process.

In conventional die casting, gases are typically vented from the die. However, the amount of gas that must vent from the dies is much greater than that of just the die cavity. All gases in the runner system must be vented as well as any volume of the shot sleeve not filled with metal. When examining the volume of gas that must be evacuated from the die combined with the short cycle times of conventional die casting, one finds that it is virtually impossible for all gases to exit the die before die fill is complete, as presented in the following sample calculation.

Example Calculation 3.2

Using the conventional die casting process, a component is manufactured with a four-cavity die in a commercially available aluminum alloy. Each casting weighs 0.5 kg. The runner system with all four components attached weighs 4 kg and includes a 5.0-cm-thick biscuit that is 7.5 cm in diameter. The length of the shot sleeve is 100 cm, and the length from the cover die end to the edge of the pore hole is 75 cm. Within the die are eight vents that are 3 cm in length and 0.1 mm in width. Given that the density of aluminum is 2.7 g/cm³, answer the following questions:

1. Assuming no porosity, what is the volume of the runner system with all four components attached?
2. What is the volume of the shot sleeve from the pour hole to the die?
3. What is the volume of the entire system, including the four component cavities, the runner system, and the shot sleeve?
4. What percent of the shot sleeve volume is filled with metal for each shot?
5. What percent of the system volume is filled with metal for each shot?
6. How much gas must be managed in this example?
7. For a 1-sec shot, how fast must the gases flow through the vents?

Solution

1. The volume of the runner system with the castings is the weight divided by the density:

$$\frac{4000 \text{ g}}{2.7 \text{ g/cm}^3} = 1481 \text{ cm}^3$$

2. The volume of the shot sleeve from the pour hole to the cover die is

$$(75 \text{ cm})\pi\, (3.75)^2 = 3313 \text{ cm}^3$$

3. The volume of the entire system, including the shot sleeve and the die cavity, is the volume calculated in question 1 without the volume of the biscuit combined with the volume calculated in question 2:

$$1481 \text{ cm}^3 - (5 \text{ cm})\pi(3.75 \text{ cm})^2 + 3313 \text{ cm}^3 = 4574 \text{ cm}^3$$

4. The shot sleeve fill percentage is

$$\frac{1481 \text{ cm}^3}{3313 \text{ cm}^3} \times 100 = 44.7\%$$

5. The system fill percentage is

$$\frac{1481 \text{ cm}^3}{4574 \text{ cm}^3} \times 100 = 32.4\%$$

6. The amount of gas that must be managed in the die is equal to the volume of the shot sleeve minus the volume of the biscuit:

$$(70 \text{ cm})\pi(3.75)^2 = 3093 \text{ cm}^3$$

7. During a 1-sec shot, 3093 cm^3 of gas must travel through the vents. The total vent area in this example is

$$8 \times 0.01 \text{ cm} \times 3 \text{ cm} = 0.24 \text{ cm}^2$$

In order for all gases to escape, the velocity of the gas through the vents must be

$$\frac{3093 \text{ cm}^3 / 0.24 \text{ cm}^2}{1 \text{ sec} = 128{,}875 \text{ cm/sec}} = 4639 \text{ km/hr}$$

In the sample calculation presented above, the estimated fill time of 1 sec is very conservative. In most cases, the time to fill a die is measured in milliseconds. However, the final calculation shows that it is virtually impossible to expel all gases from the die cavity through the vents. Usually gases are entrapped in the die and compressed into the metal. Even when the gases are compressed, the porosity can manifest as blisters at a later time. This most often occurs during heat treatment.

In some cases, conventional die casters choose not to vent gases from the die at all. In such cases, all gases must be compressed into the component. Although this may produce acceptable components for certain applications, such practices do not produce high integrity products.

Vacuum die casting is a high integrity die casting process that utilizes a controlled vacuum to extract gases from the die cavities, runner system, and shot sleeve during metal injection. This process works to minimize the entrapment of gases during die fill. When utilizing a properly sized vacuum system, 95% of all gases in the die cavity can be removed at 750 mm Hg. All vacuum die casting process parameters are consistent with those used in conventional die casting. The vacuum assists the conventional process.

During processing, the vacuum should be applied to the die as long as possible to extract as much gas as possible. Care must be taken to locate the vacuum shut-off valve at the last location to fill in the die. This location may be counterintuitive, as illustrated in Figure 3.1. As with conventional die casting, vacuum die casting exhibits atomized flow. Filling of the die cavity is not planar. If the vacuum shut-off valve is placed at the furthest point from the gate, the valve may close early in the fill process. Computer modeling may be utilized to identify proper vacuum valve placement.

Timing is critical in vacuum die casting. The vacuum should be applied just after the plunger tip closes the pour hole. If the

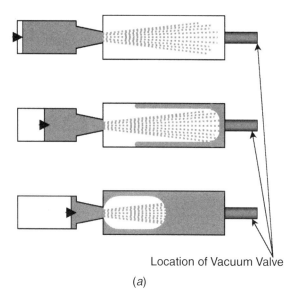

Location of Vacuum Valve
(a)

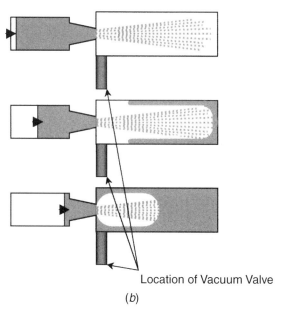

Location of Vacuum Valve
(b)

Figure 3.1 Graphical illustration showing the progression of a die cavity filling with (*a*) improper vacuum valve placement and (*b*) proper vacuum valve placement.

vacuum is applied before the pour hole is closed, air from outside the machine will be pulled through the system. If the vacuum is not applied immediately after the pour hole is closed, gases may be entrapped in the metal as waves crest and roll over within the shot sleeve. This phenomenon is illustrated in Figure 3.2.

As stated previously, gas porosity can also originate from gases dissolved in the liquid metal. Due to the short cycle times and the speed of solidification in both conventional and vacuum die casting, this form of porosity rarely occurs. Dissolved gases do not have enough time to coalesce and form porosity. Subsequent heat treating, however, can create conditions for such porosity to form into blisters. This source of porosity can be minimized by following good melting and holding practices.

3.3 MANAGING SHRINKAGE IN THE DIE

As with all die casting processes, high metal intensification pressures are maintained throughout solidification to minimize solid-

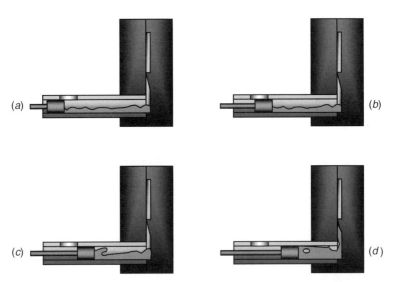

Figure 3.2 Graphical progression showing liquid metal wave cresting and gas entrapment in the shot sleeve: (*a*) pour hole open; (*b*) pour hole closed; (*c*) wave cresting; (*d*) gases trapped.

ification shrinkage porosity. However, the small gates typically used in conventional die casting freeze quickly, creating a barrier, which inhibits pressurization within the die cavity. Vacuum die casting offers no additional benefits with regards to solidification shrinkage porosity, in comparison to conventional die casting.

3.4 ELEMENTS OF VACUUM DIE CASTING MANUFACTURING EQUIPMENT

The vacuum die casting process utilizes a conventional die casting machine coupled with a vacuum system. This system is composed of a vacuum pump, a vacuum shut-off valve, a vacuum control system, and an unvented die.

Vacuum pumps must be sized in accordance with the volume of the gas, which must be evacuated from the die cavity during fill. To aid in sizing, all vacuum pumps have a characteristic operating curve that compares air flow with vacuum level. An example operating curve is presented in Figure 3.3. Vacuum pumps should be capable of pulling a vacuum of at least 725 mm Hg for use with vacuum die casting.

Rotary vane vacuum pumps are commonly used for vacuum die casting. Figure 3.4 is an illustration of such a pump. Within the

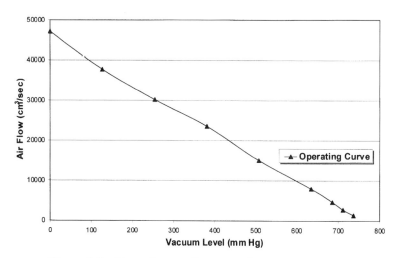

Figure 3.3 Example operating curve for a vacuum pump.

36 VACUUM DIE CASTING

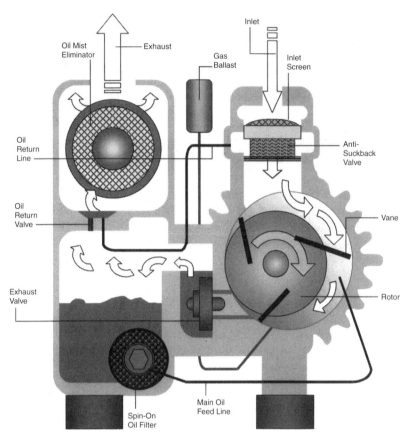

Figure 3.4 Illustration of a rotary vane vacuum pump. (Courtesy of Busch, Inc.)

pump cylinder is an eccentric rotor that pulls the vacuum. As the rotor turns, gases are trapped and compressed between several vanes and the walls of the pump cylinder. The compressed gases are discharged into the exhaust box. The gases then pass through an oil eliminator that extracts oil vapors from the exhaust gases prior to discharging them to the environment. Numerous vacuum pumps are commercially available, as shown in Figure 3.5. Often the systems are portable, making for easy placement within the die casting facility.

In vacuum die casting, a shut-off valve is needed to prevent liquid metal from entering the vacuum pump. A runner is used to

3.4 VACUUM DIE CASTING MANUFACTURING EQUIPMENT 37

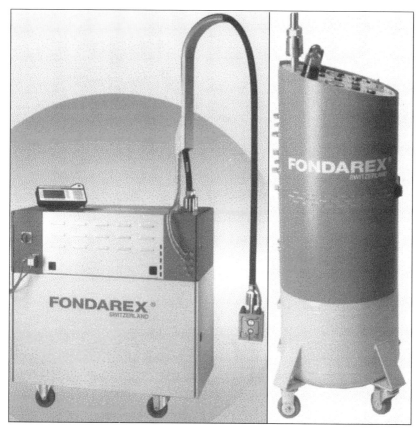

Figure 3.5 Examples of portable vacuum systems for use in vacuum die casting. (Courtesy of Fondarex Corporation.)

connect the vacuum shut-off valve to the die cavities. The gate connecting this runner to the cavity should be located at the last location in the die cavity to fill, as discussed in Section 3.2. Vacuum shut-off valves fall into two distinct categories: static and dynamic.

Static vacuum shut-off valves have no moving parts and utilize a thermal gradient to protect the vacuum system. The most common type of static vacuum shut-off valve is a corrugated chill block such as the one shown in Figure 3.6. Static vacuum shut-off valves are basically oversized vents chilled with multiple cooling lines connected to a vacuum pump. Although gases may pass through the vent, the chill solidifies liquid metal, filling the die

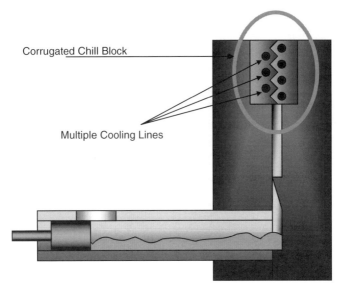

Figure 3.6 Schematic of a corrugated chill-block-type vacuum shut-off valve.

before it reaches the vacuum system. The use of a corrugated geometry forces metal to change direction numerous times as it passes through the vent. Each turn of the metal slows the fill front and aids in transferring heat into the chill. This promotes solidification and protects the vacuum pump.

Static-type shut-off valves are low in cost and are easy to maintain, as they contain no moving parts. Moreover, the vacuum pump may remain engaged and running throughout the fill, thus maximizing the amount of gases extracted from the die.

Static vacuum shut-off valves have several shortcomings. Although static shut-off valves may be easy to maintain in the die room, maintenance during production is an issue. The valve must remain clear. Flash build-up within the chill block can impede gas flow. Furthermore, a typical static vacuum shut-off vent has a width of 5–10 cm and a gap of 0.5 mm. Although a vacuum system may be capable of pulling a strong vacuum, the static valve is a bottleneck, stifling gas flow out of the die. This is illustrated in Figure 3.7. Experimental tests showed nearly a 0.5-sec lag between the vacuum pump and die cavity when using a static chill-block-type vacuum shut-off valve.

Dynamic shut-off valves offer less resistance to gas flow than static types due to their larger cross sections. Numerous designs

3.4 VACUUM DIE CASTING MANUFACTURING EQUIPMENT 39

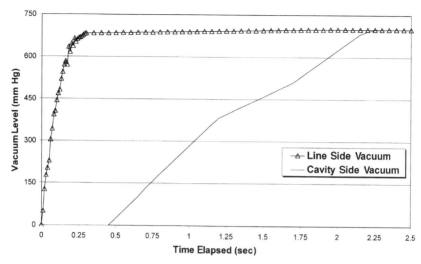

Figure 3.7 Experimental test data showing the pressure lag when using a corrugated chill-block shut-off valve (1200 cm^3 volume with 0.4 cm^2 X section). (Courtesy of IdraPrince.)

have been developed, including mechanical valves and actuated valves.

Mechanical valves are the simplest of all dynamic vacuum shut-off valves, as pictured in Figure 3.8. When the die is closed, the mechanical valve is opened. Once the pour hole is closed by the plunger tip, the vacuum is applied and evacuation of the die cavity and runner system begins. As the liquid metal fills the die, cavity pressure is used to close the shut-off valve. Mechanical valves remain open throughout cavity fill and close only after a preset cavity pressure is achieved. Closure of the mechanical valves typically occurs in 8–10 msec.

Actuated valves offer the most control in vacuum die casting as they are linked to electronic controllers. Actuated valves can either be electrical or hydraulic, as shown in Figure 3.9. The electronic control system monitors the position of the plunger. Once the pour hole is closed by the plunger tip, the electronic control system opens the valve, applying a vacuum to the die cavity. Metal enters the die and the valve is shut by the electronic controller at a preprogrammed time before metal reaches the valve. Actuated valves remain open throughout cavity fill and can be closed before metal reaches them. These valves have much larger cross-sectional areas and offer little impedance to the evacuation of gases in the

Figure 3.8 Example of a mechanical vacuum shut-off valve. (Courtesy of Fondarex Corporation.)

die cavity. This is illustrated in Figure 3.10. Electrical valves typically close in 8–10 msec; however, hydraulic valves have a much slower response time, closing in 120–150 msec.

As with any die casting process, shot control is essential. Often the shot control systems currently available on conventional die casting machines may be used to control a vacuum system.

3.5 APPLYING VACUUM DIE CASTING

Vacuum die casting builds upon conventional die casting practices by minimizing the effects of a major contributor to porosity. The cycle time and economics of vacuum die casting are equivalent to conventional die casting. The only economic penalty in using vacuum die casting is the capital cost of the vacuum system and its operation. These additional costs, however, are minor in comparison to increased component integrity.

3.5 APPLYING VACUUM DIE CASTING   41

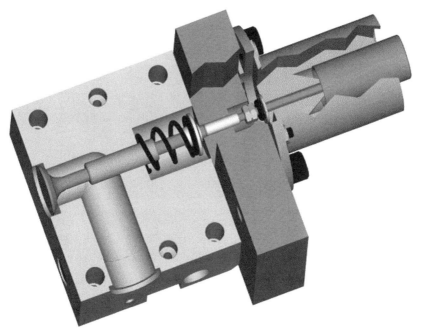

Figure 3.9 Example of a hydraulic vacuum shut-off valve. (Courtesy of IdraPrince.)

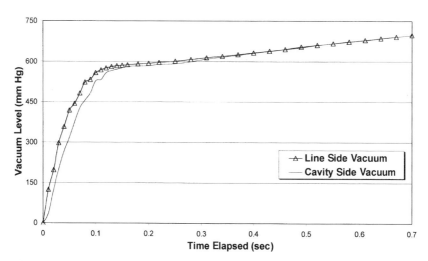

Figure 3.10 Experimental test data showing the pressure response in using a dynamic vacuum shut-off valve (1200 cm^3 volume with 1.6 cm^2 valve X section). (Courtesy of IdraPrince.)

In converting conventional die castings to vacuum die castings, one must consider the benefits that are sought. If porosity from gas entrapment is a problem, vacuum die casting can offer improvements. If shrinkage porosity is an issue, other high integrity die casting processes should be utilized.

REFERENCES

1. Sulley, L., "Die Casting," in Stefanescu, D. (editor), *Metals Handbook,* 9th ed., Vol. 15, *Casting,* ASM International, Materials Park, OH, 1988, p. 286.
2. Gordon, A., Meszaros, G., Naizer, J., and Mobley, C., *A Method for Predicting Porosity in Die Castings,* Technical Brief No. ERC/NSM-TB-91-04-C, The Ohio State University Engineering Research Center for Net Shape Manufacturing, Columbus, OH, September 1991.

CASE STUDIES: VACUUM DIE CASTING

INTRODUCTION

Over 20% of all die casting component producers in North America have vacuum die casting capabilities.[1] Each year, products numbering in the hundreds of millions are manufactured using this technology (see Figure 3.11). Although vacuum die casting is often perceived as an emerging technology, this high integrity process is decades old.

Components manufactured using vacuum die casting are utilized in a variety of applications ranging from structural members to fluid-containing vessels. When necessary, components manufactured using vacuum die casting can be heat treated without the occurrence of defects such as blistering, which commonly occurs when heat treating conventional high pressure die castings.

Vacuum die casting case studies are now presented to the design, purchasing, and manufacturing communities to illustrate the range of applications possible. The cases presented are real world achievements.

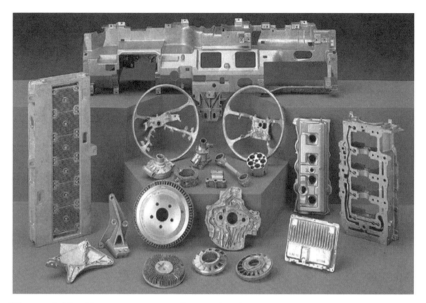

Figure 3.11 Components manufactured using vacuum die casting. (Courtesy of Gibbs Die Casting Corporation.)

B POST

Most automotive bodies are manufactured by assembling a multitude of stamped sheet metal components. However, as structural requirements have become more demanding, the number of subcomponents required in a stamped assembly has escalated. Designers have looked to castings as a means of reducing the number of required parts while meeting structural requirements. Unlike stampings, castings are net shape and can be produced with complex features and geometries. A skilled designer can integrate numerous stamped parts into a single casting.

One example of this is an automotive B post (Figure 3.12) located between a vehicle's front and rear doors. The B post is vacuum die cast in aluminum with a length of 1.2 m. The wall thickness of the casting is between 2 and 3 mm.

In order to obtain the strength required for this application, a subsequent T6 heat treatment was required. Vacuum die casting produced a part free of entrapped gases. Blistering and other de-

44 CASE STUDIES: VACUUM DIE CASTING

Figure 3.12 Automotive B post manufactured using vacuum die casting. (Courtesy of Formcast, Inc.)

fects, normally exacerbated by heat treating, were avoided. The actual material properties obtained in the finished part after heat treating included an ultimate tensile strength greater than 250 MPa and a yield strength greater than 150 MPa with approximately 15% elongation. The microstructure of the finished part had a secondary dendritic arm spacing less than 20 μm.

TRANSMISSION COVER

When considering the economics for high volume production, high pressure die casting is extremely competitive. Components traditionally manufactured using low pressure casting processes have gradually converted to high pressure die casting when possible. Vacuum technology has extended the capabilities of the die casting process, expanding the opportunities for conversions.

An aluminum transmission cover manufactured using vacuum die casting is presented in Figure 3.13. Secondary machining is required to finish several sealing surfaces of the component. Conventional die casting was unable to produce this component of an acceptable integrity such that leaking of the transmission fluid did not occur along the machined surfaces. Vacuum die casting significantly reduced the entrapped gas in the component, raising its integrity to an acceptable level for this application.

ENGINE COMPONENT MOUNTING BRACKET

Conventional high pressure die casting is rarely used to manufacture load-bearing members. Porosity from both solidification

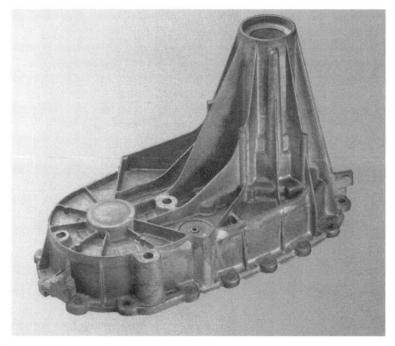

Figure 3.13 Transmission cover manufactured using vacuum die casting. (Courtesy of Gibbs Die Casting Corporation.)

shrinkage and entrapped gas significantly reduces the strength of the component. Vacuum die casting, however, can often increase the integrity of the component to an acceptable level for such applications.

Presented in Figure 3.14 is an aluminum engine component mounting bracket. Vacuum die casting was used to manufacture this product with an acceptable level of strength and ductility to survive the cyclical vibrations generated by the internal combustion engine to which it was mounted.

Vacuum die casting significantly reduced the entrapped gas in the component, raising its integrity to an acceptable level for this application.

MARINE ENGINE LOWER MOUNTING BRACKET

Product engineers are often faced with designing products that must meet a wide range of functional conditions while being faced

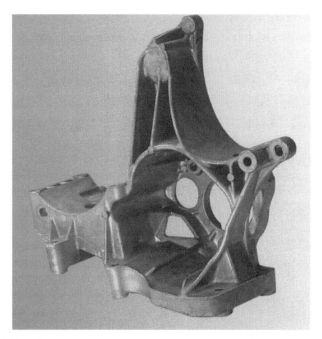

Figure 3.14 Vacuum die cast engine component mounting bracket. (Courtesy of Gibbs Die Casting Corporation.)

with economic challenges. In the case of marine engine lower mounting brackets (see Figure 3.15), impact strength was a major functional requirement. In the past, gravity permanent-mold castings with a T6 heat treatment were utilized for this application. In an effort to reduce costs, an investigation was undertaken to identify a lower cost process.

Both lost-foam casting and vacuum die casting were examined and compared to the permanent-mold casting. Although conventional high pressure die casting was a lower cost candidate than vacuum die casting, a T6 heat treatment was required. Components manufactured using the conventional high pressure die casting process often suffer from blistering when heat treated. For this reason, conventional die casting was abandoned.

To make the comparison, components were cast using each of the processes. Sections of the castings were machined into $\frac{1}{4}$-in. impact test samples. The results of the test are presented in Figure 3.16. Vacuum die castings with a subsequent T6 heat treatment were found to have the highest impact strength in comparison to the other processes. Additional testing was performed to evaluate the need for the heat treatment. Solution heat treating was found

Figure 3.15 Marine engine using a vacuum die cast lower motor mounting bracket. (Courtesy of Bombardier Motor Corporation of America.)

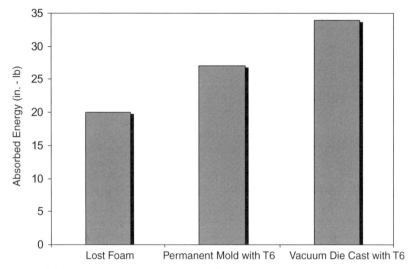

Figure 3.16 Impact properties obtained from test samples; process comparison: $\frac{1}{4}$-in. impact samples machined from castings.

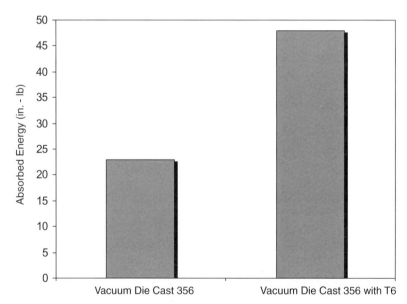

Figure 3.17 Actual impact properties obtained for the vacuum die cast marine engine lower motor mounting brackets: heat treatment benefit.

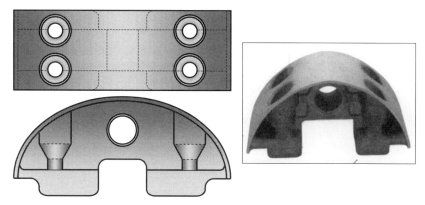

Figure 3.18 Marine engine lower motor mounting bracket manufactured using vacuum die casting. (Courtesy of Bombardier Motor Corporation of America.)

to double the impact strength of the lower motor mounting bracket, as shown in Figure 3.17.

In converting the production of the lower motor mounting bracket from permanent-mold casting to vacuum die casting, a significant cost savings was realized. The production vacuum die cast component is shown in Figure 3.18. The company that undertook this effort has chosen to apply the vacuum die casting process to several other applications as well, including engine swivel brackets and auxiliary motor mounting brackets.

REFERENCE

1. *Die Casting Industry Capabilities Directory,* North American Die Casting Association, Rosemont, IL, 2000.

4
SQUEEZE CASTING

4.1 SQUEEZE CASTING DEFINED

Porosity often limits the use of the conventional die casting process in favor of products fabricated by other means. Several efforts have successfully stretched the capabilities of conventional die casting while preserving its economic benefits. In these efforts, squeeze casting utilizes two strategies:

1. eliminating or reducing the amount of entrapped gases and
2. eliminating or reducing the amount of solidification shrinkage.

These strategies affect the major quantities that contribute to porosity as defined in Equation 1.1. These strategies will be discussed in great detail later in this chapter.

Squeeze casting is characterized by the use of a large gate area (in comparison to conventional die casting) and planar filling of the metal front within the die cavity. Squeeze casting works to minimize both solidification shrinkage and gas entrapment. Planar filling allows gases to escape from the die, as vents remain open throughout metal injection. Moreover, the large gate area allows metal intensification pressure to be maintained throughout solidification.

The origins of the squeeze casting process can be traced back to a process known as squeeze forming.[1] A schematic showing the progressive cycle of the squeeze forming process is shown in Figure 4.1. Initially, liquid metal is poured into an open die, as

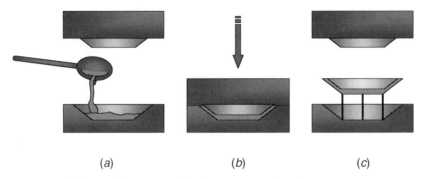

(a) (b) (c)

Figure 4.1 Schematic of the squeeze forming process.

shown in Figure 4.1a. The die is closed (Figure 4.1b) and the metal flows within the die, filling the cavity. During solidification, an intensification pressure is applied to the metal by the dies. After solidification is complete, the component is ejected, as presented in Figure 4.1c.

Squeeze casting process parameters are very similar to conventional die casting in that the liquid metal is pressurized during solidification. The major difference between squeeze casting and conventional die casting is with regards to the gate velocity. Shown in Figure 4.2 is a graph illustrating the respective process windows for numerous casting processes with respect to casting pressure and gate velocities. Gate velocities are often achieved during squeeze casting that are orders of magnitude slower than in conventional die casting. The gate velocities in squeeze casting can be as low as those characteristic to permanent-mold casting.

Cycles times for squeeze casting are longer than those of conventional die casting. This is due to both the slower metal injection speeds (required to obtain the low gate velocities noted in Figure 4.2) and the longer solidification times. The resulting microstructures are much different. Figure 4.3 compares representative microstructures for an aluminum alloy. The microstructure in the squeeze casting is not as fine as that observed in conventional die casting, and the dendrites are much more pronounced. The mechanical properties of squeeze castings are much improved due to reduced levels of porosity and the formation of microstructures not possible in conventionally die cast components. The

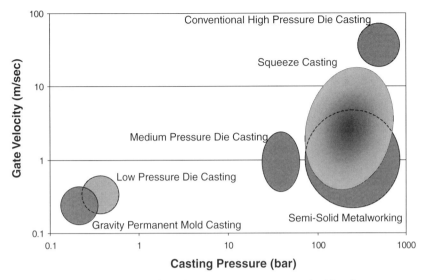

Figure 4.2 Comparisons of casting pressures to gate velocities for numerous die casting processes.

viability of subsequent heat treating is also possible due to reduced air entrapment, as discussed in the next section.

4.2 MANAGING GASES IN THE DIE

As discussed in Chapter 1, gas porosity is attributed to physical gas entrapment during die filling, to the gasification of decomposing lubricants, and to gas dissolved in the liquid alloy, which evolves during solidification. The nature of the squeeze casting process minimizes gas entrapment in comparison to conventional die casting. By utilizing larger gate cross-sectional areas and slower shot speeds, atomized fill is avoided. In many case, planar fill can be achieved during squeeze casting. This comparison is shown in Figure 4.4.

By avoiding liquid metal atomization, vents within the die remain open throughout much of cavity fill. Slower shot speeds also allow more gases to escape from the die before compression occurs.

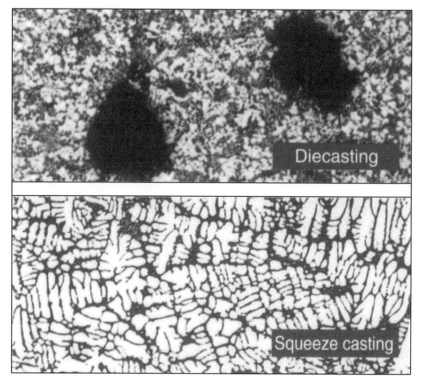

Figure 4.3 Microstructural comparisons between conventional die casting and squeeze casting. (Courtesy of UBE Machinery, Inc.)

Gas porosity can also originate from gases dissolved in the liquid metal. Although not a major factor in conventional die casting due to the extremely high cycle times, the longer solidification durations associated with squeeze casting may allow dissolved gases to precipitate and form porosity. This source of porosity can be controlled using good melting and holding practices.

4.3 MANAGING SHRINKAGE IN THE DIE

High metal intensification pressures are maintained throughout solidification in conventional and vacuum die casting. Unfortunately, the small gates typically used in conventional die casting freeze quickly. Once solidified, the gates are a barrier that inhibits further pressurization within the die.

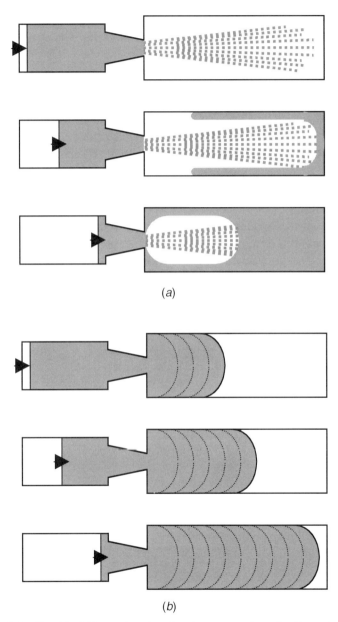

Figure 4.4 Graphical illustration showing the progression of a die cavity filling with (*a*) atomized filling and (*b*) a planar metal front.

Due to increased gate areas in comparison to conventional and vacuum die casting, gates typically remain open throughout much of component solidification when using the squeeze casting process. Pressurized metal is fed to the die cavities, reducing solidification shrinkage and minimizing the effect of this porosity-forming mechanism.

4.4 ELEMENTS OF SQUEEZE CASTING MANUFACTURING EQUIPMENT

Both horizontal and vertical conventional die casting machines can be used in conjunction with the squeeze casting process. The differences in squeeze casting are attributed to the die design and process parameters.

Although squeeze casting has been utilized for many years to manufacture production components, a consistent die design methodology has not been documented in the technical literature. Squeeze casting die design philosophies are viewed by many producers as a trade secret. As such, most producers do not wish to disclose die design methodologies. However, several qualitative characteristics are known.

In comparison to conventional die casting, squeeze casting dies have larger gate areas. Gates are no less than 3 mm in thickness to avoid premature solidification during intensification. Some manufacturers utilize classical fan gating such as that used in conventional die casting. Other producers have found large single-point gates ideal.

As squeeze castings have thicker gates, trimming is not a viable option to removing components from their runner systems. Sawing is typically required. Automated sawing systems are available for high volume production. However, automated systems require customer fixtures. Although sawing may be necessary for removing components from their respective runner systems, trimming often is not avoided. The removal of overflows and flash is still accomplished using traditional trimming techniques.

As with any die casting process, shot control is essential. Often the shot control systems currently available on conventional die casting machines may be used with the squeeze casting process. Process parameters, however, must be adjusted to allow for slower fill of the die cavity and longer intensification times. Key process characteristics of squeeze casting include metal temperature, melt cleanliness, cavity pressure, and gate velocities.[2]

4.5 APPLYING SQUEEZE CASTING

Squeeze casting is a high integrity die casting process that builds upon conventional die casting practices and is compatible with aluminum, magnesium, zinc, and copper alloy systems. Cycle times are longer in comparison to conventional die casting due to longer metal injection durations. Component integrity is improved by minimizing entrapped air and reducing solidification shrinkage. Most squeeze casting components can be heat treated without blistering defects to improve mechanical properties.

Squeeze cast components have many advantages over conventional die castings. A qualitative comparison of these two processes is shown in Figure 4.5.[3] Conventional die casting is lower

Feature	*CDC*	*Squeeze*
Metal temperature	-	-
Cycle time	+	-
Number of cavities	+	+
Alloy flexibility	+	+
Shrink porosity	-	+
Oxide entrapment	-	+
Equipment cost	++	+
Automation	++	+
Metal cost	+	+
Recycling	+	+
Mechanical properties	-	+
Heat treatable	-	+
Metal heating	+	+
Metal loss	+	+

+ = indicates favorable rating

Figure 4.5 Comparisons of conventional die casting and squeeze casting.

cost in the areas of capital equipment. Squeeze casting has additional costs associated with automated sawing for separating the runner system from squeeze cast components. A saw must be purchased, operated, and maintained along with fixturing.

These additional costs, however, are often offset with benefits in the areas of porosity reduction related to solidification shrinkage, which improves mechanical properties. Moreover, the reduction in entrapped gas results in a heat-treatable casting.

In converting conventional die castings to squeeze castings, one must consider the benefits sought. If porosity from gas entrapment and solidification is a problem, squeeze casting can offer improvements. If only entrapped gas is an issue, vacuum die casting may be sufficient. Moreover, squeeze casting can be combined with vacuum die casting. The use of a vacuum system during squeeze casting can further reduce entrapped gas beyond that normally achieved when squeeze casting. Components currently produced using higher cost manufacturing methods can be converted to squeeze casting while maintaining functional requirements. Examples of conversions are presented next.

REFERENCES

1. Dorcic, J., and S. Verma, "Squeeze Casting," in Stefanescu, D. (editor), *Metals Handbook,* 9th ed. Vol. 15, *Casting,* ASM International, Materials Park, OH, 1988, p. 323.
2. Corbit, S., and R. DasGupta, "Squeeze Cast Automotive Applications and Squeeze Cast Aluminum Alloy Properties," Paper Number 1999-01-0343, Society of Automotive Engineers, Warrendale, PA, 1999.
3. DasGupta, R., and D. Killingsworth. "Automotive Applications Using Advanced Aluminum Die Casting Processes," Paper Number 2000-01-0678, Society of Automotive Engineers, Warrendale, PA, 2000.

CASE STUDIES: SQUEEZE CASTING

INTRODUCTION

In 1997, only nine component producers were identified with squeeze casting capabilities in North America.[1] Regardless of the limited number of producers, squeeze casting process capacity has grown to an extent that millions of products are manufactured each year using this high integrity process. Examples of such components are illustrated in Figure 4.6. Products manufactured using squeeze casting are employed for a wide range of applications. Squeeze cast components produced worldwide include cross members, control arms, steering knuckles, pistons, engine mounts, and wheels, just to name a few.

Several squeeze casting case studies are now presented to illustrate real world achievements. In many cases, components manufactured using the squeeze casting process are heat treated to maximize strength and impact resistance.

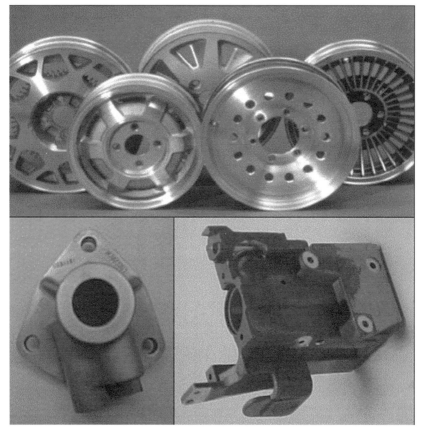

Figure 4.6 Components manufactured using the squeeze casting process. (Courtesy of SPX Contech Corporation.)

STEERING KNUCKLE

In the past, automotive steering knuckles were manufactured strictly from cast iron. With ever-increasing demands to minimize vehicle weight, the dominance of ferrous casting alloys has faltered for this application in lieu of the lower density aluminum alloys. In most cases, the squeeze casting process is preferred when converting this product from a ferrous casting to an aluminum casting. An example squeeze cast aluminum steering knuckle is illustrated in Figure 4.7.

In order to obtain the required strength for this application, a secondary T6 heat treatment was necessary. The A356 casting

Figure 4.7 Steering knuckle manufactured using the squeeze casting process. (Courtesy of Formcast, Inc.)

alloy was used in manufacturing the steering knuckle pictured in Figure 4.7. The squeeze casting process produced a component free of entrapped gases. As such, blistering and other defects, normally aggravated by the T6 heat treatment, were avoided.

The actual material properties measured in the finished part after heat treatment included an ultimate tensile strength greater than 320 MPa and a yield strength greater than 240 MPa. When tested, the heat-treated casting also exhibited an elongation of 10%. The microstructure of the finished part had a secondary dendritic arm spacing of approximately 25 μm.

VALVE HOUSING

In the past, pressure and fluid-handling products were manufactured by machining solid metal stock. In most cases, the amount of material removed during machining exceeded the weight of the final product. Half of a product's material cost ended up on the floor of the manufacturing facility as scrap. Through the use of net-shape castings, machining may be minimized, resulting in a significant economic savings.

Due to porosity from shrinkage and entrapped gas, the use of conventional high pressure die castings for fluid-handling and pressure vessel related applications are not feasible. Squeeze casting, however, is a viable means of manufacturing high integrity components suitable for fluid handling products. The aluminum valve housing presented in Figure 4.8 was manufactured using the squeeze casting process. In comparison to conventional high pressure die casting, squeeze casting significantly reduced the amount of shrinkage porosity and entrapped gas, raising its integrity to an acceptable level for this application.

Unlike valve housings machined from solid metal stock, the squeeze cast housing required the removal of only 0.136 kg of aluminum for finishing.[2] To obtain an acceptable burst strength, secondary heat treating of the squeeze casting was necessary. The strength of the heat-treated squeeze cast valve housing exceeded the burst strength requirements with an actual measurement in excess of 20.5 MPa.

62 CASE STUDIES: SQUEEZE CASTING

Figure 4.8 Squeeze cast valve housing. (Courtesy of SPX Contech Corporation.)

STEERING COLUMN HOUSING

As demands on automotive structures escalate, the functional requirements on individual components have increased. Steering column housings have been manufactured using conventional high pressure die casting for decades. Initially, the housings were manufactured using zinc alloys via the hot-chamber die casting process. In an effort to reduce vehicle weight, these housings were converted to lower density aluminum alloys manufactured using the cold-chamber die casting process. Further demands to reduce weight and improve impact strength have exceeded the capabilities of conventional high pressure die casting. As an alternative, part producers are turning to heat-treated squeeze castings as means of meeting the current requirements.

Presented in Figure 4.9 is an illustration of a squeeze cast steering column housing. Heat treatment of the housing was viable due

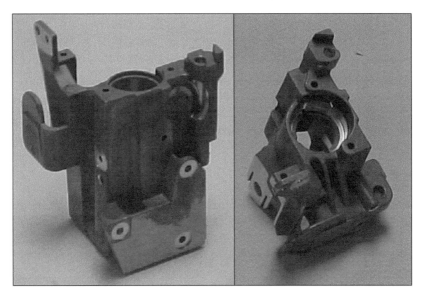

Figure 4.9 Steering column housing produced using the squeeze casting process. (Courtesy of SPX Contech Corporation.)

to the minimal amount of entrapped gas in the casting made possible by the squeeze cast process. A yield strength in excess of 280 MPa was achieved in the finished component.[2] The squeeze casting also exhibited exceptional wear resistance.

HIGH PERFORMANCE ENGINE BLOCK

As high integrity die casting technology has evolved, the size of the components produced has greatly increased. When considering the economics for high volume production, high integrity die casting processes are extremely competitive. Large products, traditionally manufactured using forging or other low pressure casting processes, have converted to die castings when possible. Automotive engine blocks are quintessential examples.

Porsche AG chose the squeeze casting process for the manufacture of the Boxter engine block, which is shown in Figure 4.10. Cast using an aluminum alloy with 9 wt % silicon and 3 wt % copper, the weight of the finished component is 20.45 kg.

Past efforts to utilize an all-aluminum engine block have failed due to wear between the pistons and the engine block. To alleviate

64 CASE STUDIES: SQUEEZE CASTING

Figure 4.10 Porsche Boxter engine block produced using the squeeze casting process. (Courtesy of UBE Machinery, Inc.)

this problem, the Boxter engine block has integrated metal matrix composite liners that are inserted into the die prior to metal injection.[3] Unlike conventional high pressure die casting, the squeeze casting process has an extended cavity filling cycle. The long filling cycle combined with high intensification pressures during solidification allows a penetrating bond to form between the liners and the casting alloy.

REFERENCES

1. *Squeeze Casting and Semi-Solid Molding Directory,* North American Die Casting Association, Rosemont, IL, 1997.
2. DasGupta, R., and D. Killingsworth. "Automotive Applications Using Advanced Aluminum Die Casting Processes," Paper Number 2000-01-0678, Society of Automotive Engineers, Warrendale, PA, 2000.
3. Merens, N. "New Players, New Technologies Broaden Scope of Activities for Squeeze Casting, SSM Advances," *Die Casting Engineer,* November/December 1999, p. 16.

5
SEMI-SOLID METALWORKING

5.1 SEMI-SOLID METALWORKING DEFINED

Semi-solid metalworking is a process in which a partially liquid–partially solid metal mixture is injected into the die cavity. Although the origins of semi-solid metalworking can be traced back over 30 years, the process did not become commercialized for high volume production until the early to mid 1990s. The process was developed at the Massachusetts Institute of Technology as an outgrowth from hot-tearing research in the 1970s.[1,2] When inducing hot tears, researchers found that partially solidified metal was thixotropic and could be deformed under pressure. With an understanding of semi-solid metal behavior, several new net-shape manufacturing processes were developed based on closed die forging, die casting, extrusions, rolling, and hybrids of these processes.[1–3]

Semi-solid metalworking has been applied to numerous metal systems, including aluminum, magnesium, zinc, titanium, and copper as well as numerous ferrous alloys. The most common commercial alloy systems in use are aluminum and magnesium die casting alloys. These alloys are ideal for use in semi-solid metalworking due to their wide freezing ranges. Examples for aluminum are shown in Table 5.1. Material that is heated to a semi-solid state can be formed, sheared, or cut easily. In Figure 5.1, a semi-solid billet of aluminum is cut with a knife.

The most common semi-solid metalworking process variants in use today parallel high pressure die casting. Shown in Figure 5.2

TABLE 5.1 Freezing Ranges for Common Die Cast Aluminum Alloys

Aluminum Alloy Designation	Approximate Solidification Range (°C)
319	604–516
356	613–557
357	613–557
380	593–538
383	582–516
390	649–507

Source: From Ref. 4.

Figure 5.1 Aluminum billet in the semi-solid state. (Courtesy of Formcast, Inc.)

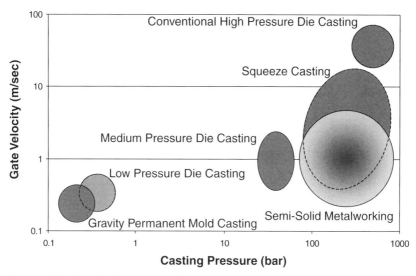

Figure 5.2 Comparisons of casting pressures and gate velocities for numerous die casting processes.

is a graph illustrating the process windows for numerous casting processes with respect to casting pressure and gate velocities. Gate velocities during semi-solid metalworking are comparable to those achieved during squeeze casting.

A controlled injection system meters a partially solid–partially liquid metal mixture into a permanent metal die to manufacture net-shape products. Semi-solid metalworking processes are more costly than traditional high pressure die casting. This is due to the equipment and energy costs associated with preparing semi-solid metal. However, the integrity of components manufactured with semi-solid technology is much improved in comparison to conventional die casting components. Semi-solid metalworking extends the capabilities of conventional die casting by

1. reducing the amount of entrapped gases,
2. reducing the amount of solidification shrinkage, and
3. modifying the microstructure of the alloy.

All strategies discussed in Chapter 1 for stretching the die casting process are addressed by semi-solid metalworking.

5.2 MANAGING GASES IN THE DIE

Gas porosity is attributed to physical gas entrapment during die filling, to the gasification of decomposing lubricants, and to gas dissolved in the liquid alloy, which evolves during solidification. Semi-solid metalworking exhibits planar metal flow due to the highly viscous behavior of semi-solid metal combined with larger gate cross-sectional areas and slower shot speeds in comparison to conventional die casting. Planar metal flow allows vents to remain open throughout much of cavity fill. Slower shot speeds also allow more gases to escape from the die before compression of the gases occurs.

Semi-solid metalworking is often incorrectly sighted as exhibiting laminar flow when filling the die cavity.[2,5,6] This misconception has proliferated in the sales and marketing of semi-solid metalworking related products. Regardless of the increased viscosity of the semi-solid metal mixture, the high flow rates encountered when filling the die cavity under production conditions results in turbulence. This turbulence, however, does not cause gases to be entrapped in the metal. Entrapment of gases occurs at the metal fill front.

5.3 MANAGING SHRINKAGE IN THE DIE

When utilizing semi-solid metalworking, a reduction in solidification shrinkage porosity is realized as a result of injecting metal that is already partially solid into the die. Also, the amount of heat, which must be removed to complete solidification, is reduced for the same reason. This allows cycle times to be shortened in comparison to high pressure die casting while simultaneously reducing the magnitude of thermal cycling to costly dies.

As with all die cast processes, high metal intensification pressures are maintained throughout solidification. Due to larger gate areas in comparison to conventional and vacuum die casting, pressurized metal is fed to the die cavities, further reducing solidification shrinkage.

5.4 MICROSTRUCTURES IN SEMI-SOLID METALWORKING

Unlike products manufactured using traditional casting methods, the microstructure of products manufactured using semi-solid metalworking is not dendritic. During processing, the dendritic structure is broken up and evolves into a spheroidal structure. The mechanical properties of the spheroidal microstructure is superior to those found in castings with dendritic microstructures as reported in numerous case studies.[1,5,7] In many cases, the strength of products produced using semi-solid metalworking rivals that of forgings.

Numerous variants of the semi-solid metalworking process exist. However, all of the processes can be grouped into one of two categories: direct processing and indirect processing. As the name implies, indirect semi-solid metalworking does not immediately produce a component. Stock material must first be manufactured with a spheroidal microstructure. This is accomplished by casting bar stock while stirring the solidifying metal mechanically or with magneto-hydrodynamic technology. The stock material is then reheated to the desired forming temperature and injected into the die while in the semi-solid state. Direct processes avoid the production and reheating of stock material. The semi-solid metal mixture is produced on demand and injected directly into the die. This greatly reduces the total cycle time. Processing cycle comparisons and microstructural comparisons are presented in Figure 5.3 between (a) direct semi-solid metalworking, (b) indirect semi-solid metalworking, and (c) conventional casting processes.

Typical as-cast microstructures for an aluminum alloy produced using the direct and indirect semi-solid metalworking are shown in Figure 5.4 and 5.5, respectively. The round white objects in the microstructures are primary aluminum spheroids, which make up the solid fraction of the material during manufacture. The surrounding matrix in the microstructure (formerly the liquid portion during manufacture) is composed of fine primary aluminum dendrites and the eutectic phase. During indirect semi-solid metalworking, liquid metal may become entrapped within the solid fraction of the material. This entrapped liquid appears as dark

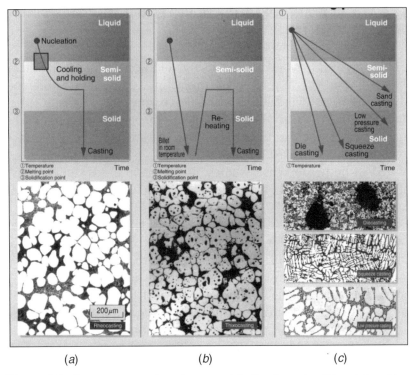

Figure 5.3 Process comparison between (*a*) direct semi-solid metalworking, (*b*) indirect semi-solid metalworking, and (*c*) conventional casting processes. (Courtesy of UBE Machinery, Inc.)

microstructural features within the spheroids, as shown in Figure 5.5.

The solid fraction of the alloy can be varied during semi-solid metalworking. Components manufactured with high solid fractions typically have finer spheroidal microstructures than components manufactured with low solid fractions.

5.5 SEMI-SOLID METALWORKING EQUIPMENT

The capital equipment and raw materials required for direct and indirect semi-solid metalworking are very different. These major

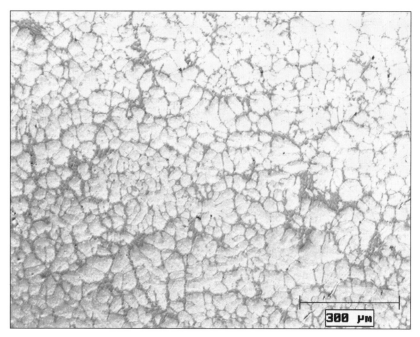

Figure 5.4 Microstructure of an aluminum component produced with a direct semi-solid metalworking process.

process variants will be discussed separately in the following sections.

5.5.1 Billet-Type Indirect Semi-Solid Metalworking

When utilizing an indirect semi-solid metalworking process, stock material with a spheroidal microstructure is heated to a semi-solid state and injected into a die. A typical indirect semi-solid metalworking manufacturing cell is pictured in Figure 5.6. Such cells begin by sawing continuously cast semi-solid stock material into billets of known volume. Billets are fed to a rotary induction heating system that raises the temperature of the billets to the semi-solid state. Automation is used to load the shot sleeve of the casting machine, which injects the semi-solid metal into the die. (Often a single robotic arm is used for loading both the induction heating system and the casting machine.) The solidified compo-

74 SEMI-SOLID METALWORKING

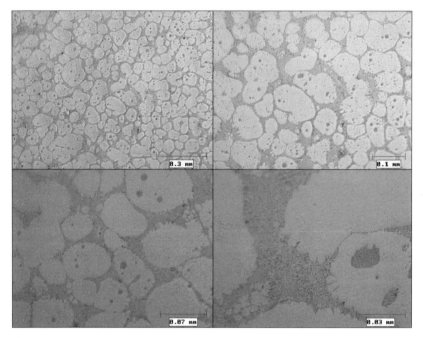

Figure 5.5 Microstructure of an aluminum component produced with an indirect semi-solid metalworking process.

nent is ejected and extracted from the die using automation that then loads the component into a saw or trim press for removal of the runner system and flash.

Systems are commercially available for sawing semi-solid feedstock into billets as well as robotic arms, conveyors, and trim presses. Induction systems for heating semi-solid billets are similar to those used by the forging industry. Induction systems used in semi-solid metalworking are typically of the rotary type with at least seven heating stations. Five of these stations are used to preheat the billet. The remaining two are used to bring the billet to the semi-solid state. Control systems in induction heating systems often include contact thermocouples. Some systems utilize a plunge thermocouple that penetrates into the heated semi-solid billet just prior to loading into the shot sleeve.

Both horizontal and vertical conventional die casting machines can be used in conjunction with indirect semi-solid metalworking. As with any die casting process, shot control is essential. Often

5.5 SEMI-SOLID METALWORKING EQUIPMENT 75

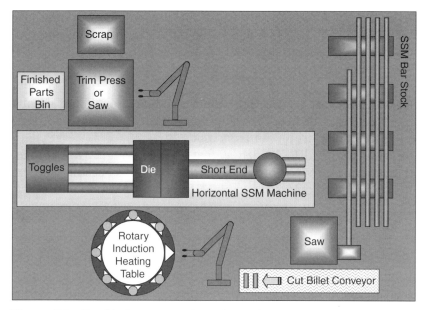

Figure 5.6 Graphical representation of a typical indirect semi-solid metalworking manufacturing cell.

the shot control systems currently available on conventional die casting machines may be used with the semi-solid metalworking process. Process parameters, however, must be adjusted to allow for slower fill of the die cavity.

Although semi-solid metalworking has been utilized for many years to manufacture production components, a consistent die design methodology has not been documented in the technical literature. Semi-solid metalworking die designs are viewed as a trade secret by most component producers. As such, very little information has been disclosed regarding die design philosophies. Nonetheless, several qualitative characteristics are known about semi-solid metalworking dies.

In comparison to conventional die casting, semi-solid metalworking dies have larger gate areas. Gates are no less than 3 mm in thickness to avoid premature solidification during intensification. Some manufacturers utilize classical fan gating such as that used in conventional die casting. Other producers have found large single-point gates ideal. Single-point gates, however, may lead to phase separation. This phenomenon is discussed in Chapter 11.

As components produced using semi-solid metalworking have thick gates, trimming is not a viable option for removing the runner systems. Sawing is typically required. Automated sawing systems with custom fixtures are available for high volume production. Although sawing is required for removing components from their respective runner systems, trimming is not avoided. The removal of overflows and flash is still accomplished using traditional trimming techniques.

Semi-solid feedstock such as that shown in Figure 5.7 is manufactured using two primary methods: continuous casting and extrusion. The structure and method of handling semi-solid billets vary depending on the way in which the feedstock is manufactured.

Magneto-hydrodynamic technology is utilized when continuously casting semi-solid feedstock as a means of stirring the liquid metal as it solidifies. This stirring action creates a fine equiaxed structure within the core of the feedstock. Regardless of the stirring induced by the electromagnetic field, a dendritic structure forms on the outer surface of the continuously cast feedstock. This dendritic case forms as a result of the substantial temperature gradient present at the liquid metal–mold interface. The dendritic

Figure 5.7 Continuously cast semi-solid metalworking feedstock. (Courtesy of Formcast, Inc.)

case is shown in Figure 5.8, which illustrates the anatomy of a continuously cast semi-solid metalworking billet before heating and after heating. After a billet is heated to the semi-solid state, the liquid fraction of the billet has a tendency to rise to the top portion of the billet as the solid fraction has a tendency to settle to the bottom. This often results in bulge at the base of heated billets. The dendritic case, however, helps support the heated billet while it is transferred to the shot sleeve.

Although the dendritic case helps hold the billet together while heating, this portion of the feed material does not flow well within the die. Methods have been developed to capture the dendritic case with the biscuit when forming a component, as shown in Figure 5.9. The plunger tip used with continuously cast billet material has a blunt point that pushes the bulk material out beyond the dendritic case. The main runner can also be used to capture the dendritic case in the biscuit if it is smaller than the diameter of the billet.

Semi-solid feedstock can also be manufactured by extruding grain refined material. This feedstock has a fine equiaxed structure without a dendritic case. This may pose a problem during heating, as shown in Figure 5.10. As the liquid fraction rises, the extruded

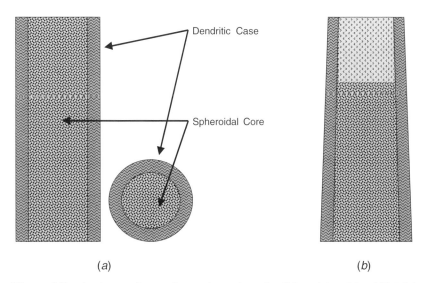

Figure 5.8 Anatomy of a continuously cast semi-solid metalworking billet (*a*) before heating and (*b*) after heating.

78 SEMI-SOLID METALWORKING

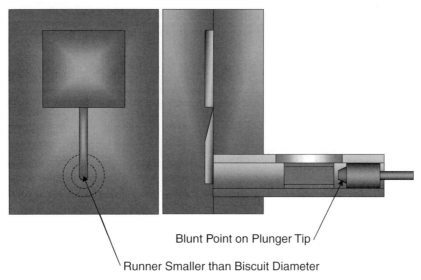

Figure 5.9 Plunger tip and die design for capturing the dendritic case of a continuously cast semi-solid metalworking billet.

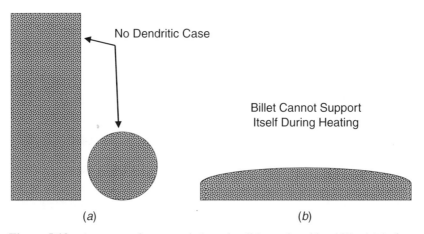

Figure 5.10 Anatomy of an extruded semi-solid metalworking billet (*a*) before heating and (*b*) after heating.

billet will sag. When utilizing extruded billet material, a crucible or other holding container must be used to contain the semi-solid charge.

5.5.2 Thixomolding® Direct Semi-Solid Metalworking

Direct semi-solid metalworking processes avoid reheating of stock material. These processes produce the semi-solid metal mixture on demand just prior to metal injection into the die. Numerous variants of this process exist. However, Thixomolding® is among the most efficient of all direct semi-solid metalworking processes, and for this reason, it will be the only direct semi-solid metalworking process discussed in this section.

A Thixomolding® machine schematic is presented in Figure 5.11. The shot end of the machine is very different from those used in all other die casting processes. A screw surrounded by heating bands, similar to those used for injection molding, is used to inject metal into the die. Moreover, the rotating action of the screw mechanically shears the heated metal creating a semi-solid mixture. Controlled quantities of pellet or chipped feedstock are metered into the screw in an inert atmosphere after each cycle.

A typical manufacturing cell is pictured in Figure 5.12. Thixomolding® begins by blowing feedstock into the feed bin mounted

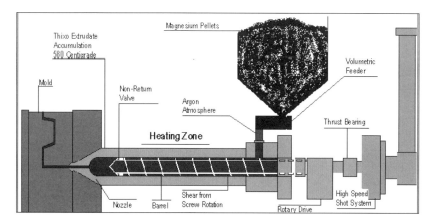

Figure 5.11 Schematic of Thixomolding® machine use in direct semi-solid metalworking. (Courtesy of Thixomat.)

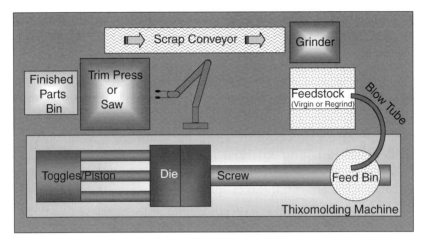

Figure 5.12 Graphical representation of a typical manufacturing cell.

above the molding machine. The feedstock is metered into the screw, heated, sheared, and injected into the die cavity. The solidified component is ejected and extracted from the die using automation that then loads the component into a saw or trim press for removal of the runner system and flash. Scrap is conveyed to a grinder, which chips the off-fall into usable feedstock.

In comparison to conventional die casting, dies used in Thixomolding® components have one major difference. After metal injection is complete, the end of the screw freezes shut. The plug that forms (Figure 5.13) keeps the semi-solid mixture from leaking out of the screw. The die must be designed to capture this plug during metal injection, as shown in Figure 5.14. At the start of metal injection, the plug is shot into a "catch" that captures the plug. The cone-shaped feature around the catch feeds the semi-solid metal into the die cavity.

The Thixomolding® process is very flexible. Since the semi-solid metal is completely contained within the screw, the liquid–solid fraction can be varied without creating complications in handling. As such, multiple microstructures may be obtained by varying the percent solid, as shown in Figure 5.15. In extreme cases, the metal may be heated to a completely liquid state. This, in essence, turns the Thixomolding® machine into a conventional die casting machine.

5.5 SEMI-SOLID METALWORKING EQUIPMENT 81

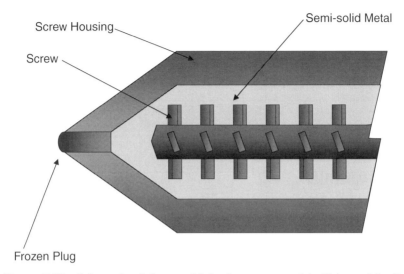

Figure 5.13 Schematic of the metal injection screw used in Thixomolding®.

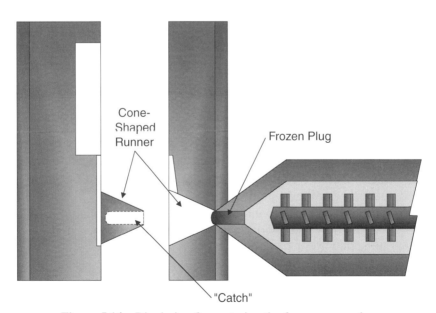

Figure 5.14 Die design for capturing the frozen screw plug.

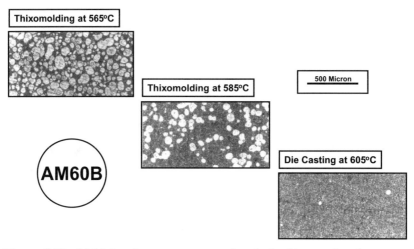

Figure 5.15 Multiple microstructures may be obtained by varying the percent solid during metal injection when using the Thixomolding® process. (Courtesy of Thixomat.)

5.6 APPLYING SEMI-SOLID METALWORKING

Semi-solid metalworking has many advantages over conventional die casting and squeeze casting. A qualitative comparison of these processes is shown in Figure 5.16.[8] Semi-solid metalworking cycle times are equivalent to those in conventional die casting. This is due to the minimal time required for solidification. Most components produced using semi-solid metalworking can be heat treated without blistering defects to further enhance mechanical properties.

Both conventional die casting and squeeze casting are lower cost in the areas of capital equipment. Unlike vacuum die casting and squeeze casting, which are used to improve the integrity of conventionally die cast components, most products manufactured using semi-solid metalworking are conversions from high cost processes, including forging and investment casting.

Although semi-solid metalworking initially was used exclusively by the aerospace industry, the automotive industry has embraced the technology. Several structural and pressure vessel applications are now in production utilizing aluminum components manufactured from semi-solid processes. Next, example alu-

Feature	CDC	Squeeze	SSM
Metal temperature	-	-	+
Cycle time	+	-	+
Number of cavities	+	+	++
Alloy flexibility	+	+	-
Shrink porosity	-	+	++
Oxide entrapment	-	+	+
Equipment cost	++	+	-
Automation	++	+	-
Metal cost	+	+	-
Recycling	+	+	- for billet processes
Mechanical properties	-	+	++
Heat treatable	-	+	++
Metal heating	+	+	-
Metal loss	+	+	- for billet processes

+ = indicates favorable rating

Figure 5.16 Comparisons between conventional die casting, semi-solid metalworking, and squeeze casting.

minum case studies are presented. The electronics industry has also found semi-solid metalworking ideal in the manufacture of numerous magnesium components. Case studies for magnesium components are presented as well.

REFERENCES

1. Flemings, M., "Behavior of Metal Alloys in the Semisolid State," *Metallurgical Transactions,* Vol. 22B, June 1991, p. 269.
2. Keeney, M., J. Courtois, R. Evans, G.Farrior, C. Kyonka, A. Koch, K. Young, "Semisolid Metal Casting and Forging," in Stefanescu, D. (editor), *Metals Handbook,* 9th ed., Vol. 15, *Casting,* ASM International, Materials Park, OH, 1988, p. 327.
3. Alexandrou, A., and G. Burgos. "Semisolid Metal Processing" in M. Tiryakioglu and J. Campbell (editors), *Materials Solutions 1998: Advances in Aluminum Casting Technology,* ASM International, Materials Park, OH, 1998, p. 23.
4. Jorstad, J., and W. Rasmussen, *Aluminum Casting Technologies,* 2nd ed., American Foundry Society, Des Plaines, IL, 1993.
5. Young, K, "Semi-solid Metal Cast Automotive Components: New Markets for Die Casting," *Transactions Number T93-131,* North American Die Casting Association, Rosemont, IL, 1993.

6. Siegert, K., and R. Leiber, "Thixoforming of Aluminum," SAE Paper Number 980456, Society of Automotive Engineers, Warrendale, PA, 1998.
7. Wolfe, R., and R. Bailey, "High Integrity Structural Aluminum Casting Process Selection," SAE Paper Number 2000-01-0760, Society of Automotive Engineers, Warrendale, PA, 2000.
8. DasGupta, R., and D. Killingsworth, "Automotive Applications Using Advanced Aluminum Die Casting Processes," SAE Paper Number 0678, Society of Automotive Engineers, Warrendale, PA, 2000.

CASE STUDIES: ALUMINUM SEMI-SOLID METALWORKING

INTRODUCTION

Currently, only eight aluminum component producers using the semi-solid metalworking process operate in North America.[1,2] However, casting machine builders and raw material suppliers are beginning to produce products with the intent to support the needed infrastructure of this emerging industry. Six machinery builders are producing casting machines compatible with semi-solid metalworking technologies. Raw material for the indirect semi-solid metalworking process is being produced in ever-increasing quantities.

Even though semi-solid metalworking is perceived as an emerging process, millions of aluminum products are manufactured each year using this high integrity process. The automotive industry is the primary user of these components. Automotive examples are illustrated in Figure 5.17. The economic benefits of semi-solid metalworking have captured the interest of other industries. Shown in Figure 5.18 are several nonautomotive examples currently in production. Several case studies are now presented of aluminum components manufactured using the semi-solid metalworking process. Each example illustrates the capabilities of the process.

FUEL RAILS

Semi-solid metalworking has the capability to produce components that are net shape. This offers the design community the

CASE STUDIES: ALUMINUM SEMI-SOLID METALWORKING   85

Figure 5.17 Automotive components manufactured using semi-solid metalworking processes. (Courtesy of Formcast, Inc.)

Figure 5.18 Components manufactured using semi-solid metalworking processes. (Courtesy of Formcast, Inc.)

86 CASE STUDIES: ALUMINUM SEMI-SOLID METALWORKING

unique opportunity to combine numerous subcomponents into a single product. Automotive fuel rails exemplify the degree of integration possible using the semi-solid metalworking process.

Fuel rails are vital to an automobile's operation by retaining and targeting fuel injectors to optimize engine performance and minimize combustion emissions. In addition, the fuel rail holds pressurized fuel, which it delivers to each injector. The majority of automotive fuel rails are manufactured by brazing together preformed tubes with numerous stamped parts. The dimensional variability of the final brazed assembly is compounded by the tolerances within each of the subcomponents. Dimensional stability is further diminished by thermal distortion inherent in the brazing process. Moreover, product integrity is suspect as each brazed joint is a potential leak path. Manufacturers of automotive fuel rails rely on 100% inspection to protect the customer from defective assemblies. Quality is not designed into the product.

In an effort to replace the high cost–low quality brazing process, many automotive engineers have turned to semi-solid metalworking to produce high integrity fuel rails. The fuel rail shown in Figure 5.19 is manufactured using the semi-solid metalworking process (bottom) replaced a brazed assembly (top) composed of 12 individual subcomponents. The aluminum component is made

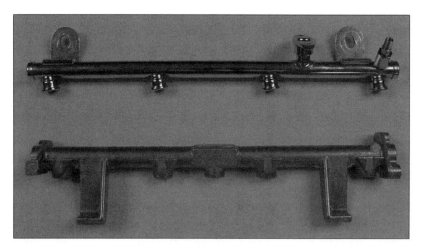

Figure 5.19 Fuel rails manufactured using semi-solid metalworking (bottom) and brazing (top).

up of one part cast near net shape with minimal secondary machining. Since no compounding of tolerances exists with the single component, improved dimensional stability is a major benefit in this application.

The fuel rail presented in Figure 5.20 is utilized on four-cylinder 2.0- and 2.2-liter passenger car engines. In this specific application, the engine configuration required a fuel rail capable of withstanding impact related to a vehicle crash. After impact, leakage of fuel was unacceptable. Traditional tubular brazed designs did not meet the impact requirements in this application. An aluminum fuel rail was designed with a 6-mm-thick wall. Manufactured using the semi-solid metalworking process, the aluminum

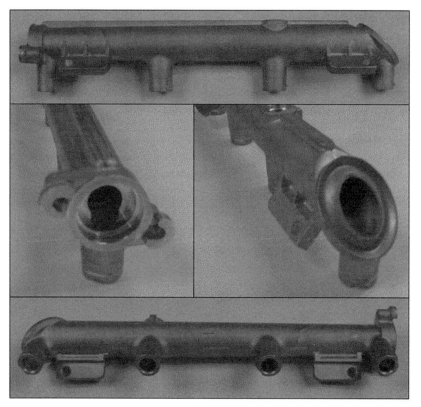

Figure 5.20 Fuel rail for use on 2.0-liter and 2.2-liter engines manufactured using an indirect semi-solid metalworking process. (Courtesy of Madison-Kipp Corporation.)

design with subsequent heat treating met the impact requirements, creating a reliable, high integrity "pressure vessel."

CONTROL ARM

Automotive control arms have typically been manufactured from ferrous castings or forgings. With ever-increasing demands to minimize vehicle weight, the dominance of ferrous components has wavered in lieu of high integrity aluminum components. A control arm (Figure 5.21) manufactured using indirect semi-solid metalworking technology is an excellent example of this trend.

In order to obtain the required strength for this application, subsequent heat treatment of the control arm was necessary. The semi-solid metalworking process utilized to manufacture this component yielded a product free of entrapped gases, making secondary heat treatment possible. The actual material properties measured in the finished component after heat treating included an ultimate tensile strength greater than 335 MPa and a yield strength greater than 280 MPa. When tested, the heat-treated component was found to have 8% elongation.

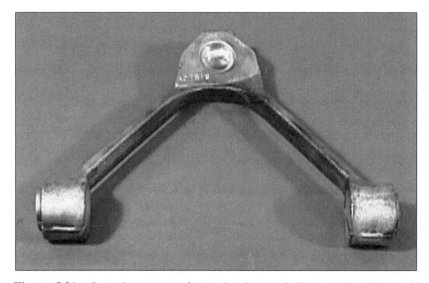

Figure 5.21 Control arm manufactured using an indirect semi-solid metalworking process. (Courtesy of Formcast, Inc.)

SWIVEL BRACKET

Product designers are often faced with the challenge of meeting a wide range of functional requirements while constrained by economics. In the case of marine outboard motor swivel brackets, impact strength was a major functional requirement. In the past, these components were manufactured using gravity permanent-mold castings with a T6 heat treatment. In an effort improve product strength and reduce costs, the swivel bracket was redesigned for the semi-solid metalworking process.

Shown in Figure 5.22 is the redesigned 7-lb swivel bracket for use on 50-horsepower outboard motors. The director of materials engineering, intimately involved in the design of the component, stated that his company "has demonstrated lower manufacturing costs and higher performance using SSM compared to semi-permanent mold swivel brackets."[2] The component manufactured with the semi-solid metalworking process required minimal machining in comparison to its semi-permanent-mold counterpart.

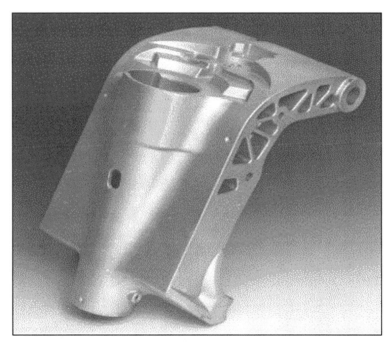

Figure 5.22 Outboard motor swivel bracket manufactured using semi-solid metalworking. (Courtesy of Formcast Inc.)

This reduction in secondary operations resulted in a significant cost savings.

IDLER HOUSING

Traditionally, automotive idler housings have been manufactured from machined iron castings that are attached to the idler arm through the use of a swaging process. This forming process bends a lip around the idler arm assembly. As such, both yield strength for part functionality and ductility for assemblability are essential requirements for this component.

In an effort to reduce weight, automotive designers looked to aluminum semi-solid metalworking technology to produce an idler housing with adequate strength and ductility. Figure 5.23 shows the completed assembly. The idler housing is manufactured in a

Figure 5.23 Automotive idler housing manufactured using an indirect semi-solid metalworking process. (Courtesy of Madison-Kipp Corporation.)

357 aluminum alloy with a T5 heat treatment. The conversion from iron to aluminum reduced component weight by 50%. The T5 heat-treated component met all material specifications, including an ultimate tensile strength of 310 MPa, a yield strength of 240 MPa, a hardness range between 83.5 and 93.5 on the "E" Rockwell hardness scale, and a minimum elongation of 7%.[3]

Several hundred thousand aluminum idler housing have been machined, assembled, and put into the field. The T5 heat treatment provided the ductility needed for the swaging operation. Moreover, all assembly test requirements were met as specified for the former iron casting, including ultimate, dynamic, and fatigue testing.

REFERENCES

1. *Squeeze Casting and Semi-Solid Molding Directory,* North American Die Casting Association, Rosemont, IL, 1997.
2. Merens, N., "New Players, New Technologies Broaden Scope of Activities for Squeeze Casting, SSM Advances," *Die Casting Engineer,* November/December 1999, p. 16.
3. Wolfe, R., and R. Bailey, "High Integrity Structural Aluminum Casting Process Selection," SAE Paper Number 2000-01-0760, Society of Automotive Engineers, Warrendale, PA, 2000.

CASE STUDIES: MAGNESIUM SEMI-SOLID METALWORKING

INTRODUCTION

Magnesium consumption has increased over 400% during the last decade.[1] Industry forecasters expect magnesium usage to climb beyond 250,000 tons by 2009. Much of this demand for magnesium is being driven by both the electronics industry and the automotive industry. These industries are committed to reducing the weight of their products while increasing functionality. The majority of the products manufactured in magnesium over the last decade were produced using die casting processes. Figure 5.24 is

92 CASE STUDIES: MAGNESIUM SEMI-SOLID METALWORKING

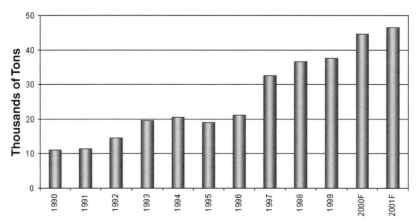

Figure 5.24 Magnesium die casting shipments in the United States, 1990–2001.

a graph showing the growth in magnesium die casting shipments within the United States since 1990.

Even though magnesium is a high cost material in comparison to steel or iron when looking strictly at cost per pound, this price differential does not take into account the lower density of magnesium. Moreover, magnesium components can be produced as net-shape products in high volumes. For many complex steel fabrications, conversion to a single net-shape magnesium component results in a significant reduction in cost with assembly and tooling while reducing weight.

Due to the volatility of magnesium while in a liquid state reacting with the atmosphere, few casting methods can process magnesium as safely and economically as die casting. Direct semi-solid metalworking of magnesium allows a component producer to process magnesium with virtually no atmospheric interaction. Numerous components are currently manufactured in magnesium using semi-solid metalworking, as shown in Figure 5.25. Several case studies are now presented of magnesium components produced using direct semi-solid metalworking processes. These examples illustrate the capabilities of this process, which continues to evolve at a rapid pace.

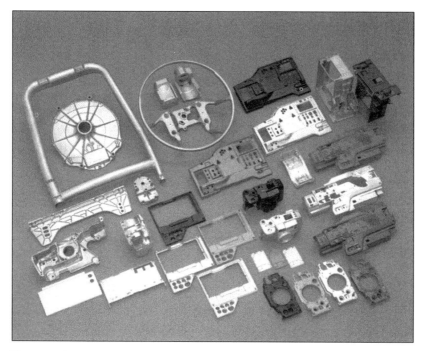

Figure 5.25 Magnesium components manufactured using semi-solid metalworking processes. (Courtesy of Thixomat.)

AUTOMOTIVE SEAT FRAME

Traditionally, automotive interior structures have been fabricated in steel by welding together stampings and formed tubes. With the continual drive to economically reduce weight and improve quality, automotive designers have turned to magnesium castings as means of manufacturing these components in one piece. Many automotive interior structures are now being produced in cast magnesium including instrument panel structures and seat frames, as shown in Figure 5.26.

Formerly produced in numerous steel stampings, the prototype magnesium seat frame shown in Figure 5.26 was manufactured using direct semi-solid metalworking technology. The magnesium seat frame measures 0.46 m in width by 0.57 m in height. A 35% weight reduction was achieved by manufacturing the component

94 CASE STUDIES: MAGNESIUM SEMI-SOLID METALWORKING

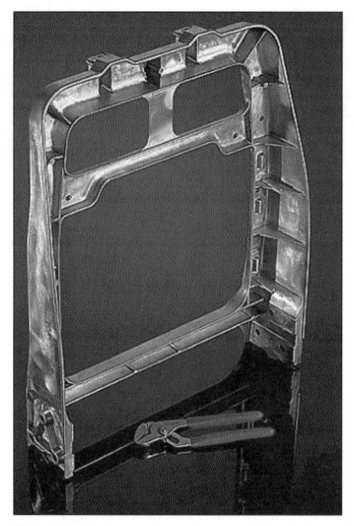

Figure 5.26 Magnesium automotive seat frame manufactured using semi-solid metalworking. (Courtesy of Thixomat.)

in magnesium in comparison to its steel counterpart. Also, part count was reduced due the net-shape capabilities of the semi-solid metalworking process. Ribbing and other structurally reinforcing geometric features were incorporated into the magnesium design.

WIRELESS TELEPHONE FACE PLATES

In the past, the cases and housings for most portable electronic products have been manufactured in plastic materials using the injection molding process. Subsequent shielding was required within the housing to protect the encased components from externally generated electromagnetic fields. With consumer demands to reduce weight, the electronics industry has turned to magnesium for these housing applications. The magnesium performs a dual role by producing net-shape geometrically complex housings while simultaneously shielding sensitive electronics from electromagnetic interference. In many cases, these magnesium components are manufactured using direct semi-solid metal working processes.

Among the top users of magnesium within the electronics industry are wireless telephone manufacturers. Figure 5.27 is an

Figure 5.27 Wireless telephone face plates manufactured in magnesium using semi-solid metalworking. (Courtesy of Thixomat.)

illustration of magnesium face plates for wireless telephones manufactured using semi-solid metalworking. The wall thickness of the components ranges from 0.8 to 1.0 mm with an overall weight of 105 g.[2] These components are produced to a net-shape geometry with minimal finishing required. Unlike many of the engineered plastics used in the past, the magnesium face plates are 100% recyclable, providing an added bonus.

VIDEO PROJECTOR CASE

From an electronic industry design standpoint, magnesium offers many functional benefits over competing materials. These benefits include low density, excellent stiffness, electromagnetic shielding capabilities, heat-sink properties, and net-shape manufacturability. By utilizing semi-solid metalworking processes in the manufacture of geometrically complex magnesium components, wall thicknesses between 0.5 and 2.0 mm can be obtained with excellent dimensional repeatability.

An example of these capabilities is shown in the liquid crystal display (LCD) projector case illustrated in Figure 5.28. The pro-

Figure 5.28 Magnesium LCD projector case manufactured in three sections using semi-solid metalworking. (Courtesy of Thixomat.)

jector case consists of three pieces: the base, cover, and front panel. Each component was manufactured using the Thixomolding® process with virtually no secondary finishing operations required, other than tapping. Integrated into the base are numerous bosses and ribs. Due to the thermal conductivity of magnesium, the need for large internal fans, previously needed to cool components, was eliminated. This resulted in a more compact design with an overall weight reduction of 40% in comparison to the previous generation manufactured using a stamped steel frame. Also, the net-shape magnesium case reduced assembly time by 50% by eliminating the need for separate hardware and fasteners. Specifically, two sets of 2.39-mm (+0.038- to −0.025 mm) diameter locating pins spaced 178 mm apart are held within ±0.051 mm. Also, two 2.5-mm-diameter locating holes are cast within a ±0.025-mm diametric tolerance. These features reduced the overall part count as well as the need for secondary machining.

CAMERA HOUSING

Portability has become the number one functional requirement for many consumer electronic products. In most cases, portability equates to weight. As with wireless telephones and portable projectors, hand-held electronic cameras have turned to magnesium in an effort to reduce weight and satisfy customer demands.

The camera housing pictured in Figure 5.29 is composed of four separate components. Manufactured using the Thixomolding® process, these complex net-shape magnesium components are produced with a quality metallic finish requiring no subsequent machining. In comparison to engineered thermoplastics, the rigidity and impact resistance of the housing are improved. Moreover, the housing shields electromagnetic interference without the need for coatings, platings, or additional parts.

LAPTOP COMPUTER CASE

Personal computers have evolved at unprecedented rates over the last decade. In many cases, personal computers have moved from

98 CASE STUDIES: MAGNESIUM SEMI-SOLID METALWORKING

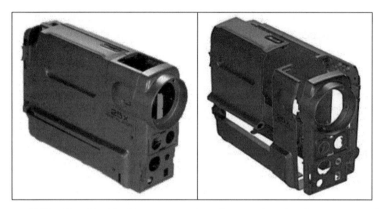

Figure 5.29 Hand-held video camera housing produced using a magnesium semi-solid metalworking process. (Courtesy of Thixomat.)

the desk to the briefcase, as is the case for laptop computers. In this competitive industry, computer users demand light weight.

In order to meet these demands, manufacturers have come to rely on magnesium components when producing compact laptop computers. Furthermore, the electronics industry is taking advantage of the benefits intrinsic to semi-solid metalworking processes. Shown in Figure 5.30 is a compact computer case manufactured in magnesium using a direct semi-solid metalworking process. The case is lighter and stronger than the plastic materials formerly used for this application. The result is an attractive net-shape case with excellent thermal conductivity and inherent shielding to electromagnetic interference.

POWER HAND TOOL HOUSING

The use of magnesium semi-solid metalworking is not limited to the automotive and electronics industries. Other consumer industries have found this high integrity process ideal for several applications.

Shown in Figure 5.31 is a power hand tool housing manufactured in magnesium using semi-solid metalworking technology. This component illustrates the flexibility in design possible with this high integrity die casting process. The dimensional variability

CASE STUDIES: MAGNESIUM SEMI-SOLID METALWORKING 99

Figure 5.30 Compact laptop personal computer case manufactured using semi-solid metalworking in lightweight magnesium. (Courtesy of Thixomat.)

100   CASE STUDIES: MAGNESIUM SEMI-SOLID METALWORKING

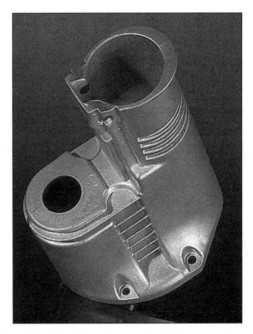

Figure 5.31   Power hand tool housing produced in magnesium using semi-solid metalworking. (Courtesy of Thixomat.)

within the manufactured housing is limited, allowing a bearing to be pressed into a net-shape hole without the need for subsequent machining. The finished part is lightweight and impact resistant, essential requirements of the fully assembled product.

REFERENCES

1. *Die Casting Industry Capabilities Directory,* North American Die Casting Association, Rosemont, IL, 2000.
2. M. Lessiter (editor), "Castings in Action," *Engineered Casting Solutions*, Winter 1999, p.55.

6
THERMAL BALANCING AND POWDER DIE LUBRICANT PROCESSES

6.1 THERMAL CYCLING INHERENT TO HIGH INTEGRITY DIE CASTING PROCESSES

Inherent in all high integrity die casting processes is the thermal cycling of tooling. Every cycle exposes the surface of the die to extreme temperature changes, as shown in Figure 6.1.[1] Initially metal enters the tool, raising the die surface temperature near the melting point of the alloy being cast almost instantaneously. The metal in the die quickly begins to cool along with the die surface. When the metal solidifies, the heat of fusion liberated slows the cooling of the die surface. The solid metal and the surface of the die continue to cool. When the die is opened and the solid component is ejected, the die surface will continue to cool until an equilibrium temperature is reached with the bulk mass of the die. Typically, liquid lubricants are sprayed onto the surface of the die between metal injections. This thermally shocks the die surface as well. Although this may cool the surface of the die quickly, once the spray is stopped, the surface temperature will rise back to the temperature of the bulk die mass.

The aggressive thermal cycling that occurs on the die surface can lead to many defects. Moreover, the typical containment strategies to counteract these defects often lead to other problems. Discussed in this chapter are the defects, current methods to con-

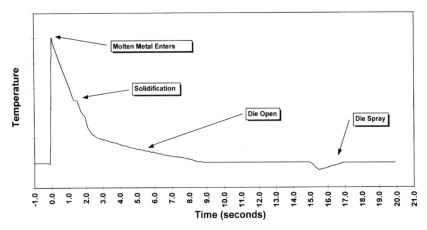

Figure 6.1 Die cavity surface temperature over a single processing cycle.

tain such defects, and alternative strategies to combat these problems for use with high integrity die casting processes.

6.2 HEAT CHECKING AND SOLDERING

Die surfaces are typically hardened tool steel with a martensitic microstructure. This hard wear-resistant surface is excellent for use with high integrity die casting processes. However, martensite is also brittle. The aggressive thermal cycling inherent in high integrity die casting processes often causes the surface to crack. Over time initial microcracks will develop into larger cracks, or heat checks, penetrating into the die. This progression is shown in Figure 6.2. Once cracks form on the die surface, liquid metal injected into the die will fill these voids. The result is checking and small veins on the surface of the casting.

In many cases, minor heat checking is not a problem. However, for components in which surface quality is a requirement, heat checking is a costly defect often requiring the addition of secondary finishing and polishing operations. Regardless of surface finish requirements, heat checking is the dominant mode of failure for dies as product dimensional stability will eventually be affected.

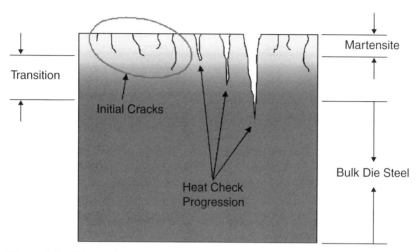

Figure 6.2 Die surface cross section illustrating heat-checking formation and progression.

Soldering is another problem, which can occur when utilizing high integrity die casting processes. This phenomenon occurs when a bond forms between a component and the die surface during solidification, often resulting in permanent damage to both the component and the die surface. Soldering typically occurs at hot spots on the die surface. Examples of such locations include small uncooled cores and points subjected to impinging metal flow. Areas around welded die repairs are also prone to soldering. The repair process often transforms the resistant martensitic layer on the die face into a soft annealed surface (Figure 6.3) These annealed areas are much more susceptible to soldering.

6.3 CONTAINING THE EFFECTS OF HEAT CHECKING AND SOLDERING

Several strategies are used by component producers to contain heat checking and soldering. In all cases, economics are affected.

Since many components produced using high integrity die casting processes do not have stringent surface finish requirements, low cost shot blasting may be used to remove heat checking as

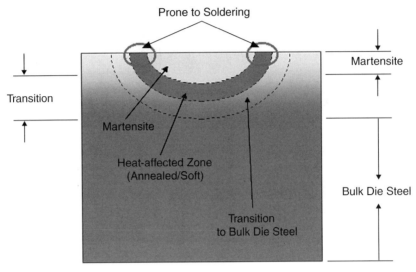

Figure 6.3 Die surface cross section illustrating microstructural weaknesses in a welded die surface.

well as flash. For components with stringent surface requirements, dies are often designed for easy maintenance. Die cavities are often replaced when heat checking occurs. This leaves the majority of the die untouched. Even severe heat checking of the runner system does not affect component functionality. The most costly solution to resolve heat checking is the complete replacement of a die.

To combat soldering, many strategies may be used. In some cases, a slower cycle time may eliminate the problem, allowing water lines to cool the die. More often, component producers look to liquid die surface lubricants. Longer spray times may be used to apply additional lubricant across the entire die surface. This may overlubricate the die. An alternative is to spray additional lubricant on select areas of the die prone to soldering. Component metal chemistry may also be adjusted. Additions of iron can reduce the soldering potential of an alloy. However, iron additions have a negative effect on mechanical properties. In many cases, once soldering occurs, it cannot be totally eliminated. Replacement of die cavity inserts may be the only solution in these cases.

6.4 REPERCUSSION OF HEAT CHECKING AND SOLDERING CONTAINMENT ACTIONS

Containment of heat checking and soldering problems does have repercussions. The most obvious is cost. The replacement of die inserts is costly. Beyond the cost of the tooling, resources must be expended to measure and qualify tooling for dimensional accuracy and functionability. Longer spray times increase cycle times. This reduces productivity and increases materials cost through the use of additional lubricant.

In some cases, the containment actions for combating soldering induce other defects. Spraying hot spots on the die surface to resolve soldering problems induces additional thermal cycling and heat checking. Additional lubrication on the die surface may combust and form porosity in components during processing.

The root cause of heat checking is thermal cycling. The root cause of soldering is variation in die surface temperature during processing. All of the strategies suggested contain the problems of heat checking and soldering. None address the root cause.

6.5 THERMAL MANAGEMENT OF HIGH INTEGRITY DIE CASTING PROCESS TOOLING

Although most component producers work to optimize die cooling to eliminate hot spots and accelerate solidification, few work to minimize variation in die surface temperature. Minimization of die surface temperature variation, or thermal balancing, can resolve several processing issues.

Cooling lines typically account for 90% of all heat removal from a die during processing. Numerous tools are available to aid in thermal balancing and die surface temperature control. Computer models have evolved to a point where cooling line placement can be optimized before dies are built.[2] Infrared cameras are a noncontact method of measuring die temperature. Some infrared cameras are capable of examining entire die faces for easy analysis of temperature variation. Closed-loop coolant temperature control units are also commercially available to better control heat flow out of the die.

Thermal balancing offers component producers several benefits, including

1. a reduced potential for solder,
2. elimination of thermal shock and heat checking at hot spots,
3. reduced cycle time from shorter spray cycles, and
4. less lubricant usage.

Although thermal balancing is a way to control several potential problems encountered in high integrity die casting processes, commitment to upfront engineering is a necessity.

6.6 MINIMIZATION OF THERMAL CYCLING EFFECTS WITH POWDER LUBRICANTS

Powder lubricants are one method of minimizing the magnitude of thermal cycling in high integrity die casting processes. Instead of applying a lubricant with a liquid carrier, the lubricant is applied dry. This eliminates the thermal shock caused by spraying a room temperature liquid on the die face. Powder lubricant methods, however, rely entirely on cooling lines to remove heat from the die surface. Thermal balancing is a necessity when using powder lubricants.

Successful application of powder lubricants to the die surface is not intuitively obvious. Traditional open die spray methods produce unacceptable levels of particulate matter in the air. To minimize this environmental issue, experiments have been performed in which the powder lubricant is electrostatically charged. Although electrostatic methods reduce particulate matter in the air, overspray is a major issue. The best results have been obtained with a novel closed die process.

Application of a powder die lubricant with a closed die requires the use of a vacuum and modification of the shot sleeve for a

6.6 THERMAL CYCLING EFFECTS WITH POWDER LUBRICANTS

powder feed nozzle, as shown in Figure 6.4. The same system used in vacuum die casting can be utilized with powder lubricant application, although vacuum valve placement may not be common.

Figure 6.5 is an illustration of the powder lubricant casting cycle. Initially, the plunger is positioned in the shot sleeve (*a*) such that the pour hole is closed. A vacuum is applied to the closed die cavity while a controlled quantity of powder lubricant is metered (*b*) into the shot sleeve. The vacuum pulls the lubricant into the die cavity (*c*), creating a thin layer of lubricant on the surface. The plunger is retracted (*d*) opening the pour hole. Liquid metal is metered into the shot sleeve (*e*) and then injected (*f*) through a runner system into a die cavity (*g*) under high pressure. High pressures are maintained on the alloy during solidification. After complete solidification, the die opens (*h*) and the component is ejected.

Although powder die lubricants are commercially available, their use is not widespread. Some companies have reported extended die life, shorter cycle times, and improved plant cleanliness. The increase in productivity is attributed to the elimination of spray automation.

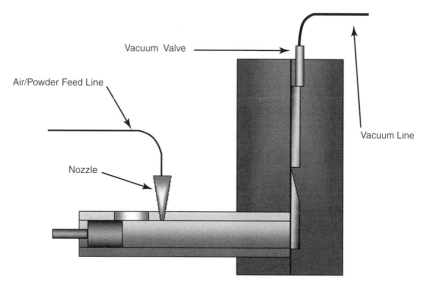

Figure 6.4 Shot sleeve and tool design for use with closed die powder lubricant application.

108 THERMAL BALANCING AND POWDER DIE LUBRICANT PROCESSES

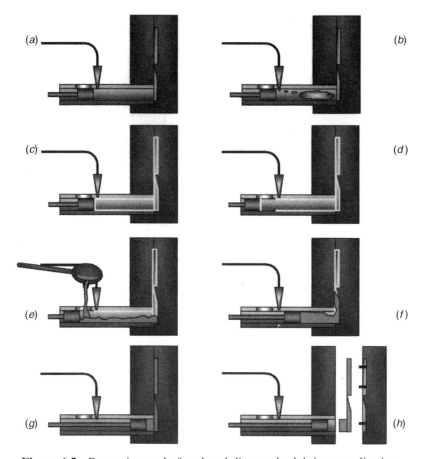

Figure 6.5 Processing cycle for closed die powder lubricant application.

6.7 APPLYING THERMAL MANAGEMENT METHODS IN REAL WORLD APPLICATIONS

As with all high integrity die casting processes, thermal balancing of dies and powder lubricant processes require the commitment of additional resources in comparison to conventional die casting. As such, strict thermal management and balancing of dies are not required for all applications. For cases in which component surface finish and extended die life are a necessity, thermal balancing and powder lubricant processing offer opportunities for improvement.

REFERENCES

1. Vinarcik, E., "Finite Element Analysis of Process Data Curves for Statistical Process Control," SAE Paper Number 970081, Society of Automotive Engineers, Warrendale, PA, 1997.
2. Roshan, H., V. Sastri, and R. Agarwal, "Die Temperature Control in Pressure Die Casting," *AFS Transactions,* American Foundrymen's Society, Des Plaines, IL, 1991, p. 493.

DESIGN CONSIDERATIONS FOR HIGH INTEGRITY DIE CASTINGS

7
DESIGN FOR MANUFACTURABILITY OF HIGH INTEGRITY DIE CASTINGS

7.1 INTRODUCTION TO DESIGN FOR MANUFACTURABILITY

Product cost, manufacturability, and quality are intimately related to product design. A skilled manufacturer may be able to reduce scrap and optimize a process to achieve near-ideal efficiency, but the cost and quality of a product can never be improved further without changing the design. Major reductions in product cost can only be attained through discerning design work.

All manufacturing methods have limitations, high integrity die casting processes included. Skilled designers and product engineers will adjust and refine designs to take advantage of the production process chosen. This is known as design for manufacturability.

7.2 HIGH INTEGRITY DIE CASTING DESIGN FOR MANUFACTURABILITY GUIDELINES

Although high integrity die casting processes may be able to produce products that cannot be manufactured using conventional die casting methods, the same design for manufacturability guidelines apply for optimizing product geometry. This criterion is well documented and includes the following design considerations:

1. maintaining a consistent wall thickness;
2. using gradual transitions from surface to surface;
3. eliminating large metal masses;
4. using comers, fillets, and radii to assist with metal flow;
5. using ribs to facilitate metal flow; and
6. maintaining a sufficient draft angle.[1,2]

By following these six general guidelines, the majority of manufacturing problems caused by the design can be avoided.

7.3 AUTOMOTIVE FUEL RAIL CASE STUDY REVIEW

The purpose of this case study is to review the designs of two fuel rails manufactured using the semi-solid metalworking process shown in Figure 7.1. For discussion purposes, the fuel rails will be designated Z-1 (top) and Z-2 (bottom). Both fuel rails are intended for use in four-cylinder engines of equal size. After intro-

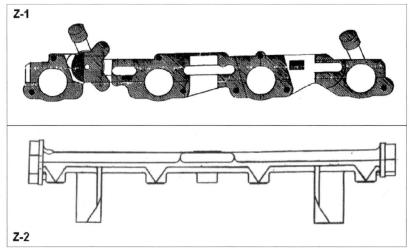

Figure 7.1 Production fuel rails Z-1 (top) and Z-2 (bottom).

ducing fuel rail functional requirements and the analysis methods used in this study, observations are presented related to the manufacturability of the two designs with respect to die filling, leak tightness, and dimensional stability.

7.3.1 Fuel Rail Functional Requirements

Components used in automotive fuel systems must maintain high structural integrity and pressure tightness to assure customer safety and meet ever-increasing durability and performance requirements. Over the past decade, the increased use of single port fuel injection has led to the introduction of "fuel rails" as a subcomponent in engine fuel system design. Fuel rails have evolved to perform the following functions:

1. retaining fuel injectors mounted on the top of the engine,
2. targeting injectors to optimize engine performance,
3. supporting regulators and sensors used by the fuel system, and
4. delivering fuel to each injector.

Several processes have been utilized to successfully manufacture fuel rails for the automotive industry, including brazed tubular steel, aluminum extrusions, injection moldings, and aluminum forgings. These processes were chosen due to their consistency in producing leak-tight products.

To reduce cost and improve quality, alternative processes are being considered by the automotive industry. This has led to the successful manufacture of aluminum fuel rails by several companies using high integrity die casting processes.[3-5]

7.3.2 Case Study Analysis Method

Three characteristics related to manufacturability were examined in the analysis of the two fuel rail designs:

1. complete die fill during semi-solid processing,
2. leakage due to porosity, and
3. dimensional stability.

All data related to complete die fill were collected by visual examination of fuel rails after manufacture. Leakage due to porosity was detected using air pressure decay methods after machining. Further analysis of porosity was performed by using metallographic methods. Dimensional stability was determined using coordinate measuring machines.

7.3.3 Review of the Z-1 Fuel Rail Design

Figure 7.2 is a diagram of design Z-1. The main features in this design are the fuel inlet (A) and fuel outlet (B) tubes, the fuel injector pockets (C, D, E, and F), and the fuel pressure regulator boss (G).

The Z-1 design was initially manufactured by forging an aluminum billet followed by extensive machining. The original weight of the aluminum forging billet for the Z-1 fuel rail is 1500 g. After forging, the rough fuel rail weighs roughly 800 g and is then reduced by machining to approximately 500 g. During processing, two-thirds of the weight from the initial aluminum forging billet is turned into scrap.

As a cost savings effort, production of the rough Z-1 fuel rail was shifted from forging to a high integrity die casting produced

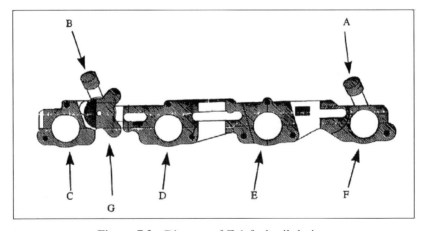

Figure 7.2 Diagram of Z-1 fuel rail design.

using the semi-solid metalworking process. This shift was made without redesigning the Z-1 fuel rail. The semi-solid rough part weighs roughly 550 g as cast and about 400 g after machining. The scrap weight and finished part weight reductions were obtained by the increased net-shape capability of semi-solid metalworking processing over forging.

The Z-1 fuel rail, as stated earlier, was originally designed to be manufactured using the forging process. Initial trials showed that the Z-1 fuel rail could be manufactured using the semi-solid metalworking process without redesign by using a gate at the center of the rail with a large cross section.

Several areas of the Z-1 design, however, were prone to fill problems, including the inlet and outlet tubes (features A and B, respectively in Figure 7.2) and the corners of the regulator boss (feature G in Figure 7.2). The tips of the injector pockets (features C, D, E, and F in Figure 7.2) also exhibited filling problems. In order to feed the inlet tube, the metal flow had to turn its direction about 120°. Fill problems such as these were detected by visual examination during production.

Shrinkage often occurred in the large masses found in the Z-1 design. Machining periodically cut into the shrinkage porosity, creating leak paths. Such issues were detected using air pressure decay methods. The fuel pressure regulator pocket, feature G in Figure 7.2, was most prone to shrinkage. The micrograph in Figure 7.3 shows shrinkage typical to that found in the fuel pressure regulator pocket.

The Z-1 fuel rail design is in conflict with many of the design for manufacturability guidelines defined for high integrity die casting processes, including

1. lack of consistent wall thickness,
2. several large metal masses,
3. sharp transitions between features, and
4. minimal use of fillets and radii.

The draft angle used in the manufacture of the Z-1 design was found to be acceptable.

118 MANUFACTURABILITY OF HIGH INTEGRITY DIE CASTINGS

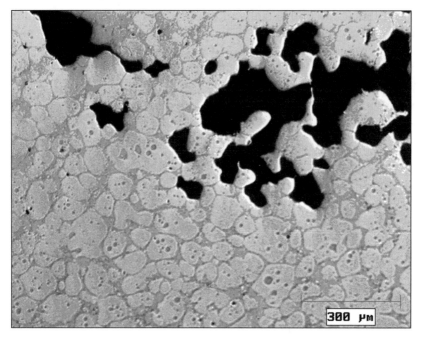

Figure 7.3 Shrinkage porosity found in the regulator pocket of the Z-1 fuel rail.

7.3.4 Review of the Z-2 Fuel Rail Design

The Z-2 fuel rail is the next generation of the Z-1 design. Figure 7.4 is a diagram of the Z-2 design with inlet (A) and outlet (B) flanges, injector pockets (C, D, E, and F), machining fixturing

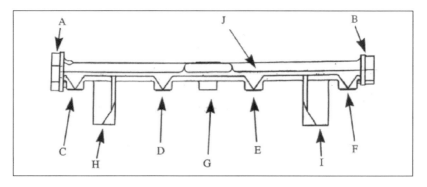

Figure 7.4 Diagram of Z-2 fuel rail design.

boss (G), mounting legs (H and I), and strengthening rib (J). As with the previous Z-1 design, a large gate located at the center was used during the manufacture of the Z-2 fuel rail.

The greatest opportunity to reduce cost in the Z-1 design was to minimize machining. The injector pocket geometry for the Z-1 design accepted two O-rings and required the machining of 11 separate features in each injector pocket. By using a different fuel injector, the Z-2 design needed to accept only one O-ring and required the machining of just 3 features in each injector pocket. Also, the Z-2 injector cup design filled much easier during semi-solid metalworking than the preceding Z-1 design.

When developing the Z-2 fuel rail, the high integrity die casting design for manufacturability guidelines presented in Section 7.2 were used with the intent to create an optimized geometry for manufacture by the semi-solid metalworking process. The inlet and outlet locations were moved to the ends of the fuel rail. This eliminated the 120° turn required for filling the inlet tube on the Z-1 design and allowed the Z-2 fuel rail to be cored from end to end. This eliminated a gun drilling operation required with the final Z-1 design. The metal mass of the as-cast part was also reduced. The shrinkage-prone fuel pressure regulator pocket used on the Z-1 design was eliminated, and a flange-mounted regulator was utilized in its place.

Several new features were added to the Z-2 design, including a machining fixturing boss, two mounting legs, and a strengthening rib. The thickness of the strengthening rib and mounting legs are uniform. The weight of the machined Z-2 fuel rail was approximately 200 g, half that of the Z-1 design. The number of machined features in the Z-2 design is 28 as compared to 95 machined features required with the Z-1 design.

Early in the development of the Z-2 design, several design configurations were developed for the end flanges. Figure 7.5 is a series of diagrams showing flange design evolution.

Figure 7.5*a* shows the initial flange design developed for ease in parting the die. This flange design, however, exhibited filling problems due to its angularity and asymmetry.

Figure 7.5*b* shows the second iteration in flange design, which improved filling by creating symmetry and increasing volume. The large volume of this design, however, resulted in shrinkage

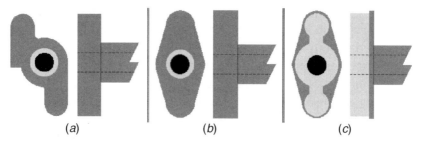

Figure 7.5 Z-2 end-flange evolution: (*a*) initial asymmetric design; (*b*) symmetric design; (*c*) symmetric webbed design.

porosity, which created leak paths after machining. Also, two metal fill fronts converged in the flange. Due to the planar fill associated with the semi-solid metalworking process, a contaminant vein formed at these converging metal fronts (see Section 11.4.1). Machining at the end flange often cut into the contaminant vein, creating leak paths. Figure 7.6 is a micrograph showing typ-

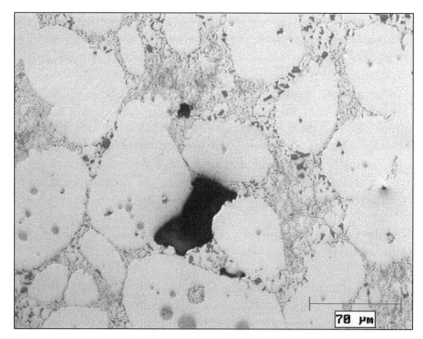

Figure 7.6 Shrinkage porosity and contaminant vein oxide inclusions found in the second flange design of Z-2 fuel rail.

ical shrinkage porosity and contaminant vein oxide inclusions observed in the second flange design.

The final design, used for production, is shown in Figure 7.5c. This design maintains symmetry for metal flow during filling and reduces the metal thickness, which in turn minimizes shrinkage. The interface between the fuel rail body and the flange also has a fillet added to assist metal flow into the flange. Although not understood at the time the design was developed, the two converging metal fill fronts in the final design are stretched and broken up by the flange ribbing. The turbulent flow created by this flange geometry dispersed oxides and contaminants on the surface of the metal fill fronts, avoiding the contaminant vein observed in the prior flange design.

When beginning production of the Z-2 design, dimensional variation was observed relative to the position of the ends. After some investigation, it was found that the strengthening rib, located on only one side of the rail, cooled at a different rate than the bulk of the fuel rail. This asymmetric cooling caused the fuel rail to bow.

7.3.5 Further Design for Manufacturability Improvements

Currently, the Z-2 fuel rail is being manufactured with a scrap rate an order of magnitude below that of the Z-1 design. Based on the above discussion, however, several improvements may be made to the Z-2 design to further increase its manufacturability. A single strengthening rib is found to distort the fuel rail during cooling. A second rib on the opposite side of the fuel rail body may be added to create a symmetric part and reduce distortion.

Although the mounting legs and the strengthening rib of the Z-2 fuel rail are of a consistent wall thickness, the thickness of the fuel rail body is not uniform. The wall varies in thickness relative to the taper of the core, which stretches the length of the fuel rail. Due to the length of the core, the wall thickness at one end of the fuel rail body is several millimeters thicker than the opposite end. Figure 7.7 is a micrograph showing shrinkage porosity in the thickest portion of the fuel rail body. (Although shrinkage porosity is present in the bulk mass of the component, the porosity is not a defect as the fuel rail remains leak tight and meets structural

122 MANUFACTURABILITY OF HIGH INTEGRITY DIE CASTINGS

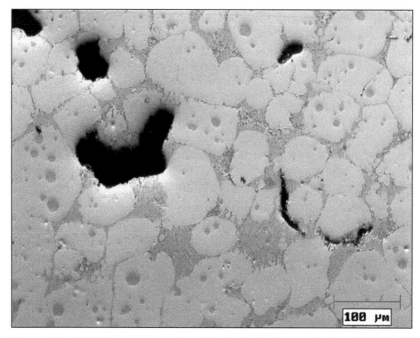

Figure 7.7 Shrinkage porosity found in the body of a production Z-2 fuel rail.

requirements.) To enhance the manufacturability of the Z-2 fuel rail design further, a second improvement would be to taper the outside of the fuel rail body to follow the taper of the center core. Following the taper of the core would create a uniform wall thickness for the body of the fuel rail.

A third area for further improvement is in the feeding of metal into the mounting legs. The current Z-2 design forces the metal to turn 90° to its natural flow path. The addition of a large fillet at the interface between the mounting leg and the body of the fuel rail would assist in metal filling.

7.4 CONCLUSIONS OF THE CASE STUDY

Observations made in this case study reinforce the design for manufacturability guidelines for high integrity die casting processes presented in Section 7.2. The two fuel rail designs reviewed in the case study functionally serve the same purpose. However, the ease in production using the semi-solid metalworking process varies greatly due to their respective geometries. Although the desire may exist to implement a high integrity die casting processes with any product, problems will always be encountered when attempting to manufacture a design that was optimized for another manufacturing method.

Six general design for manufacturability guidelines were reviewed in this chapter. The topic of high integrity die casting design, however, extends well beyond the brief discussion presented in this book. Texts are available focusing on this single subject.[6,7] In all cases, designers and product engineers should consult the manufacturing engineers and die makers ultimately responsible with manufacturing a component in production. The vested interest and feedback of these individuals will improve component quality and reduce total costs.

REFERENCES

1. Sully, L., "Die Casting," in Stefanescu, D. (editor), *Metals Handbook,* 9th ed. Vol. 15, *Casting,* ASM International, Materials Park, OH, 1988, pp. 286–295.
2. Ruden, T., *Fundamentals of Die Casting Design,* Society of Manufacturing Engineers, Dearborn, MI, 1995, pp. 19–37.
3. Moschini, R., "Manufacture of Automotive Components by Pressure Die Casting in the Semi-Liquid State," *Die Casting World,* October 1992, pp. 72–76.
4. Moschini, R., "Mass Production of Fuel Rails by Pressure Die Casting in the Semi-Liquid State," *Metallurgical Science and Technology,* Vol. 12, No. 2, 1996, pp. 55–59.
5. Jerichow, U., J. Brevick, and T. Altan, *A Review of the Development of Semi-Solid Metal Casting Processes,* Report No. ERC/NSM-C-95-45, The Ohio State University Engineering Research Center for Net Shape Manufacturing, Columbus, OH, 1995.
6. *Product Design for Die Casting,* 5th ed., Die Casting Development Council of the North American Die Casting Association, Rosemont, IL, 1998.
7. Jorstad, J., and W. Rasmussen, *Aluminum Casting Technologies,* 2nd ed., American Foundry Society, Des Plaines, IL, 1993.

8
COMPONENT INTEGRATION USING HIGH INTEGRITY DIE CASTING PROCESSES

8.1 INTRODUCTION TO COMPONENT INTEGRATION

Product cost is a function of design. A skilled manufacturer may be able to reduce scrap and optimize a process to achieve near-ideal efficiency, but the cost of a product can never be reduced further without improving the design. Major reductions in product cost can only be attained through discerning design work.

During design, the best way to minimize cost is to keep the design simple by first minimizing the number of individual components and then assuring that the remaining components are easy to manufacture and assemble. Engineers and designers must be aware that every time two parts are integrated into one, at least one operation or process is eliminated during manufacturing. In most cases several operations are eliminated along with the support activities associated with each component, such as inventory control.

8.2 HIDDEN COSTS IN EVERY COMPONENT

Every component has a cost associated with it. Often the cost is quantified by looking at the price of the raw material, but this quantification is oversimplified and leaves out much of the true financial burden. From a total-cost standpoint, the life cycle of a component is as follows:

Designed
Drafted
Quoted
Sourced to an internal or external supplier
Tooled
Approved
Manufactured
Packaged
Inventoried
Shipped
Received
Handled
Assembled into the final product

Although all of these points contribute to the total cost of a component, the attachment of a price tag to each point is difficult to ascertain. One major automotive company has estimated the administrative burden to maintain one part at $50,000. This may seem unreasonable, but the estimate includes the time to prepare and detail a drawing, approval of the design, distribution of the detailed print, cataloging the part into a worldwide database, scanning the part's drawing into a global computer site, tracking changes or updates to the drawing, and much more.

Once a product is no longer being manufactured, the financial burden continues. One must consider the servicing of products in the field. Service parts must be packaged and warehoused, sometimes for years, before they are shipped to customers.

From a quality standpoint, every component adds risk. Fewer parts means fewer things can go wrong during manufacturing, and fewer things can go wrong once the product is in the field. Every quality issue or problem carries a financial burden whether it is fixed or not.

When determining the cost of a product, one must remember to look beyond the material cost. The total financial burden of every component is made up of much more.

8.3 ANALYZING INTEGRATION POTENTIAL

Analysis methods have been developed to assist engineers and designers in the evaluation of products to determine if components may be integrated.[1] The application of these methods is often referred to as design for assembly or integrated design.

Presented in Figure 8.1 is a flow chart that summarizes the main design for assembly principles.[2] The flow chart addresses three basic areas that influence component integration:

movement for function,
material type for function, and
service.

A designer or product engineer can utilize this flow chart as a guide in analyzing an assembly to determine if component integration is possible.

8.4 COMPONENT INTEGRATION USING HIGH INTEGRITY DIE CASTING PROCESSES

Choosing a manufacturing process often is the next hurdle to component integration. Many manufacturing processes are limited in their ability to produce complex geometries. Some processes require the use of many individual parts that must be assembled into the final component. Other processes require costly secondary operations. In order to realize the benefits of integration, a flexible process is required.

Few manufacturing methods offer the flexibility obtained by using high integrity die casting processes. All high integrity die casting processes produce components that are near net shape and offer engineers the ability to go to a finished component in some-

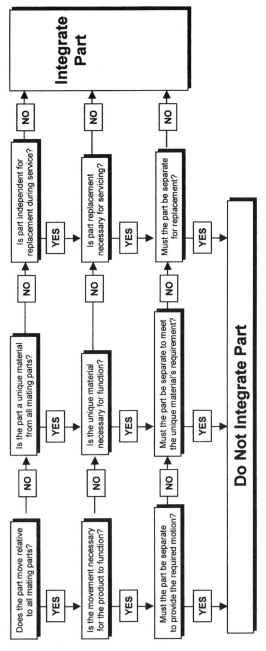

Figure 8.1 Component integration analysis flow chart.

128

times as little as one step. Extremely complex component geometries can be cast in one piece. As a result, secondary operations such as machining may be minimized or eliminated entirely.

8.5 COMPONENT INTEGRATION CASE STUDY

Automotive fuel systems offer an excellent example of part integration using a high integrity die casting process. Many automotive fuel system components are manufactured by brazing together numerous preformed tubes and stampings. Shown in Figure 8.2 is a brazed fuel rail composed of 15 individual components. The

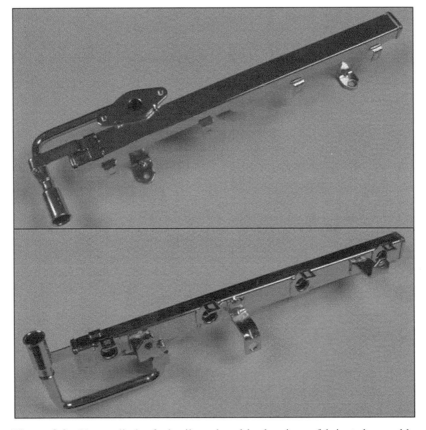

Figure 8.2 Four-cylinder fuel rail produced by brazing a fabricated assembly.

130 COMPONENT INTEGRATION USING HIGH INTEGRITY PROCESSES

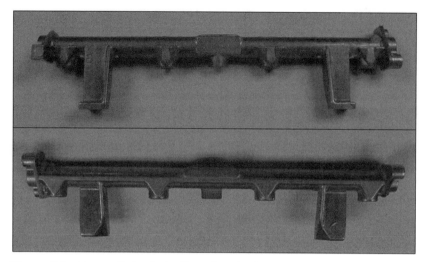

Figure 8.3 Four-cylinder fuel rail produced using the semi-solid metalworking process.

brazed assembly is also plated for external corrosion protection. The dimensional variability of the final brazed component is compounded by the tolerances within each of the individual components making up the final product.

Automotive fuel rails are also manufactured using several high integrity die casting processes, including semi-solid metalworking and squeeze casting. An example fuel rail manufactured using semi-solid metalworking is shown in Figure 8.3. Fuel rails manufactured using high integrity die casting processes are typically made up of one part cast near net shape with minimal secondary machining. Since no compounding of tolerances exists when there is one part, improved dimensional control is a major benefit in this application.

REFERENCES

1. Boothroyd, G., and P. Dewhurst, *Product Design for Assembly,* Boothroyd Dewhurst, Wakefield, RI, 1991.
2. Vinarcik, E., "Minimizing Cost through Part Integration," *Engineered Casting Solutions,* Winter 1999, p. 56.

9
VALUE ADDED SIMULATIONS OF HIGH INTEGRITY DIE CASTING PROCESSES

9.1 INTRODUCTION TO APPLIED COMPUTER SIMULATIONS

All products follow nearly the same sequential steps in development; specifically

1. research and development,
2. concept or advanced engineering,
3. product design,
4. process design,
5. process planning,
6. product launch, and
7. production.

The number of uncontrollable variables or noise factors that may cause problems to occur increases as a product moves out of the research laboratories through the stages of product development into the hands of the customer. At the same time, the number of factors that may be used to control or solve problems declines. This relationship between problem potential and avenues for problem solution is shown graphically in Figure 9.1.[1]

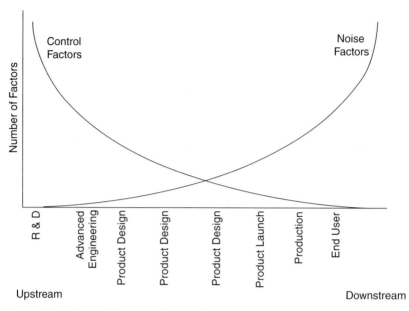

Figure 9.1 Control factors and potential problems in the product development cycle.

The cost to solve problems also increases as a product matures. As more concept and design decisions are made, resources are consumed. This is shown graphically in Figure 9.2 with the life cycle cost lever.[2] As a product approaches launch, resources have been expended to build tooling and test processes. Changes at this stage of a product's life cycle require many steps to be repeated, often with increased cost due to overtime in a desire to meet timing plans.

Figure 9.2 illustrates the benefits of predicting and solving problems early in a product's life cycle. Addressing problems late in development or after launch is a drain on resources and lowers a company's competitiveness.

Don Clausing presents three levels of competence when addressing problems during a product's life cycle[3]:

1. Problems are found. Wishful thinking allows many to be swept downstream. A large number end up in the marketplace.

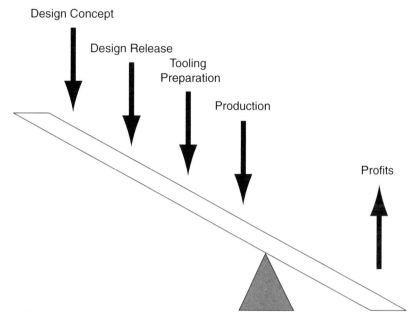

Figure 9.2 Project cost lever illustrating returns as a function of when an investment is made.

2. Problems are found. The total quality approach is used to find and correct the root causes of the problem. The information is fed upstream so that the same problem is not introduced in a later development program.
3. Problems are prevented. Potential problems and their root causes are identified before they occur. Optimization positions the design as far as possible from all potential problems. The information is fed downstream to ensure that the problem prevention decisions are understood and maintained to avoid the inadvertent later introduction of the problem.

Most companies strive for the third level of competence, in which problems are prevented. Rarely is this third level ever met. In many cases, this is because problems cannot be predicated.

Over the last decade vast improvements in computer hardware and software technology have made complex simulations of physical phenomena possible. Today engineers and designers have available an ever-growing and continually refined set of tools to aid in product development. Mathematical models using both finite element and finite difference techniques have been developed to simulate various product functions, conditions, and environments, including

- steady state and dynamic fluid flow,
- stresses in flexing structures,
- vibration and fatigue life,
- electric circuits and dynamic magnetic fields,
- thermodynamics, and
- corrosion life.

These simulations offer the design community the opportunity to predict problems early in a product's life cycle. Corrective actions often can be made to resolve problems before designs and tooling have been finalized. In some cases, mathematical models have advanced to a level in which complete product validation is possible through computer simulation, which avoids the need for costly prototypes.

9.2 COMPUTER SIMULATIONS OF HIGH INTEGRITY DIE CASTING PROCESSES

Advances also have been made in simulating metal casting processes. Specific to high integrity die casting, mathematical models have been developed to simulate several elements of the process, including

- die filling,
- air entrapment,
- liquid metal surface tracking (for predicting inclusion locations),
- solidification thermodynamics,

material properties after solidification,
shrinkage porosity, and
part distortion.

Today's computer simulations are highly developed, producing complex graphics showing the progress of metal flow during die filling (Figure 9.3). Thermodynamic results obtained from computer simulations can be used to predict numerous issues, including hot spots on the die surface prone to wear and heat checking. Such data can be used to predict regions in a component prone to solidification shrinkage, as shown in Figure 9.4. Recent advances in computer modeling of high integrity die casting processes have focused on the prediction of residual stresses and component distortion, as shown in Figures 9.5 and 9.6.

Coupling process modeling with design simulations can yield significant returns on investment by optimizing both the manufac-

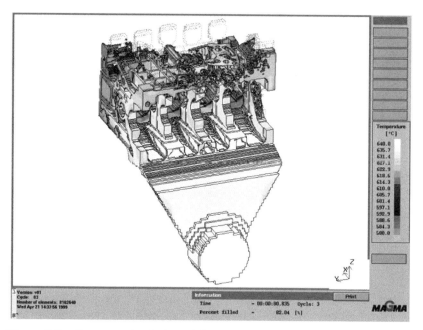

Figure 9.3 Computer simulation of die filling during metal injection. (Courtesy of MAGMA Foundry Technologies, Inc.)

136 VALUE ADDED SIMULATIONS OF DIE CASTING PROCESSES

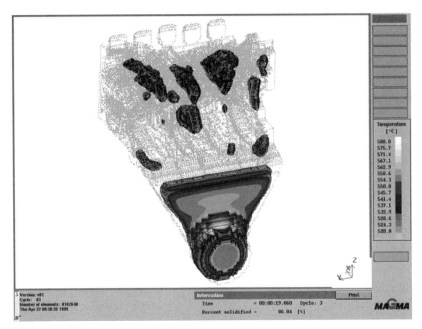

Figure 9.4 Computer simulation illustrating areas in the die cavity prone to solidification shrinkage porosity. (Courtesy of MAGMA Foundry Technologies, Inc.)

turing process and product function prior to investments in production tooling.

9.3 APPLYING SIMULATIONS EFFECTIVELY

Although some organizations report great successes when using simulations, many organizations find little value in computer modeling. Organizations can be qualitatively categorized with the respective effectiveness of their simulation efforts, as shown in Figure 9.7.[4] Although few, some organizations let their simulation capabilities sit idle while others do not use simulations to any effect, generating nothing more than pretty pictures.

The majority of simulation users fall into the two middle categories. Many organizations utilize computer modeling after designs are completed, released, and on their way to production.

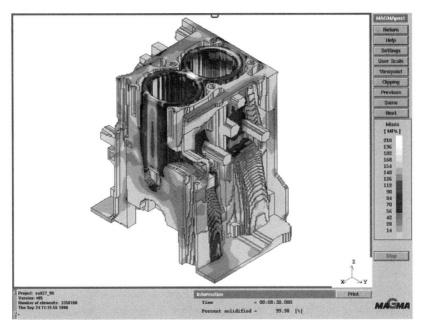

Figure 9.5 Computer simulation showing variation in residual stress that forms during solidification and cooling. (Courtesy of MAGMA Foundry Technologies, Inc.)

Under these circumstances, the opportunity is lost to optimize designs. Satisfactory and even mediocre designs are carried through to launch unless the simulation identifies a major problem. Correcting problems this late in the design cycle is costly, often requiring complete product redesign and a rush to retool. Another large group of users complete simulations and correct major problems prior to release but do little to optimize the design.

Organizations that benefit the most from computer modeling use simulations to optimize designs and in some cases validate the design prior to tooling a product for production. A benefit to using simulations in this manner is improved product quality by design as opposed to quality by inspection.

With such varying degrees of effectiveness, the common element leading to successful simulation has little to do with computer models. Effective simulations are the result of good management requiring resources and up-front planning.

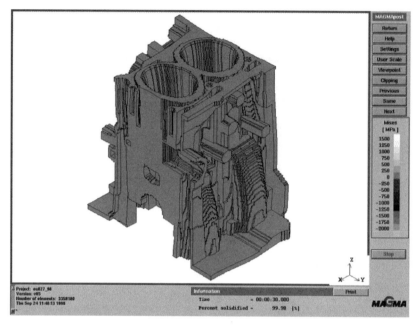

Figure 9.6 Computer simulation showing component distortion (exaggerated) during cooling. (Courtesy of MAGMA Foundry Technologies, Inc.)

9.3.1 Resources

Many organizations hold the perception that computer modeling is easy, fast, and simple requiring nothing more than the touch of a button. Often the assignment to "computer model" is given to someone in addition to a full load of other duties. This approach typically generates mediocre results at best. Computer modeling is not a substitute for engineering. Diligence must be taken in preparing the inputs to a simulation. Although a simulation may run to completion in only a few hours, the input data may take weeks to prepare. The common request by organizations to do modeling when an engineer has spare time is unreasonable. An organization must commit focused manpower if simulations are to be effective.

When committing manpower, care must be taken to select the right people. A diverse skill set is required of an individual who is to conduct any given simulation. The individual must be trained

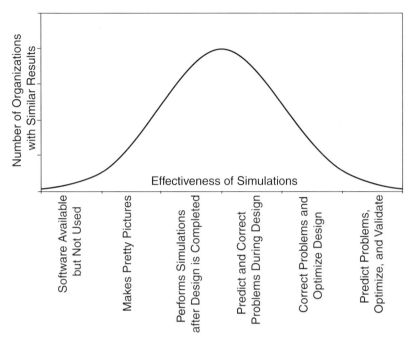

Figure 9.7 Qualitative illustration grouping the number of organizations to the effectiveness of their simulation efforts.

in modeling as well as possess knowledge of the product's function and the processes used in manufacturing. The individual must possess soft skills, as interaction is necessary between the design and manufacturing communities.

9.3.2 Planning

Effective use of computer simulations requires planning. Few organizations, if any, have the resources to conduct simulations on every product. Planning efforts should begin by identifying strategic products. For these key products, the functions and processes that are to be simulated must be defined. Depending on the complexity, multiple models may need to be run for each specific function and process. For example, a structural component may need to have simulations run to model both stresses and fatigue life. An additional set of simulations may be run by the high integrity die casting supplier to optimize the process for the component. For each individual simulation, a goal must be defined as well as a metric to measure success.

Most product timelines clearly define the date a design must be completed with fully dimensioned drawings and specification. When preparing a timeline, it is important to specify an earlier date at which a design is to be ready for use in computer simulations. Multiple iterations of any given simulation are often necessary when seeking an optimized design. Also included should be a time before design release to repeat simulations.

9.3.3 Coupling Product and Process Simulations

Often process simulations are conducted by suppliers after designs have been completed and formally sourced. In such cases, optimization of the manufacturing process is limited to process parameters. The opportunity to modify the design to improve manufacturability is lost.

Sharing design data with suppliers prior to design release with the purpose of conducting process simulation can result in significant improvements in manufacturability and reductions in cost during production. Formal drawings with complete dimensions are not required to complete such activities. Raw three-dimensional computer-aided design (CAD) data are all that is required. In many cases, results from the process simulation can be used to predict functional properties of the product as well.

9.4 COMMITMENT

When choosing to conduct simulations, an organization must be committed in order to be effective. Capable individuals must be selected and dedicated to the task of computer modeling. Simulations must be conducted prior to the completion and release of a design. An organization must plan for success.

9.5 A CASE FOR SHARING SIMULATION DATA ACROSS ORGANIZATIONS

During the development of a structural component for an automotive chassis, a supplier and original equipment manufacturer

(OEM) worked concurrently to engineer both the product and the process. Simulations were conducted by the OEM to analyze the stresses within the component during operation. The supplier used computer modeling to develop a gate and runner system. The results of the simulations were not exchanged due to "proprietary" reasons. Conclusions drawn from the results, however, were shared freely, resulting in minor changes to the design to improve function and manufacturability. At release, the partners believed both the product and process had been optimized.

Prototypes were fabricated from production, such as tooling, for product verification tests. The initial prototypes failed. Although simulations were conducted concurrently, the simulation results were never examined side by side. Comparison of the simulation results showed that two major metal fronts in the die converged at the point in the structure that bore the highest stresses during operation. The process tooling was redesigned such that the two metal fronts converged at a low stress point within the structure. Had the results of the product and process simulation been compared earlier, redesign of the process would have been avoided.

REFERENCES

1. *Robust Design Using Taguchi Methods Workshop Manual,* American Supplier Institute, Livonia, MI, 1998.
2. *Solutions,* American Supplier Institute, Livonia, MI, 1999.
3. Clausing, D., *Total Quality Development,* ASME Press, New York, 1994.
4. Vinarcik, E., "Minimizing Cost Through Part Integration," *Engineered Casting Solutions,* Winter 1999, p. 56.

CONTROLLING QUALITY IN HIGH INTEGRITY DIE CASTING PROCESSES

10
APPLYING STATISTICAL PROCESS CONTROL TO HIGH INTEGRITY DIE CASTING PROCESSES

10.1 INTRODUCTION TO STATISTICAL PROCESS CONTROL

Before the industrial revolution, craftsmen controlled the quality of their products. As the industrial revolution eased into its second century, the mass production of products by unskilled labor demanded a means to control quality. By bringing together the disciplines of engineering, statistics, and economics, statistical process control (SPC) theory was first developed in the 1920s by Walter Shewhart while working for the Western Electric Company.[1]

At the time, Western Electric was a leading manufacturer of telephones in the United States, with over 40,000 people working at the Hawthorne Works in Chicago, Illinois. Product reliability was a major problem. Manufacturing inconsistencies caused excessive failures of handsets, switching units, and amplifiers. One out of eight employees at the Hawthorne facility was an inspector whose role was to scrap defective product. Western Electric's management realized that they had a bigger problem than scrapping defective products. They were paying 12% of their workers to do nothing that added value to their products. Western Electric's management looked to minimize the sources of defective production and selected Shewhart to lead the effort.

Shewhart began studying several manufacturing processes in depth. As a practical engineer, he understood that in real life situations, laws and theories are not exact. Shewhart concluded that all processes exhibit variation that can be classified into two distinct types:

1. inherent or common-cause variation and
2. intermittent or special-cause variation.

To concisely demonstrate the difference between these types of variation, a dartboard may be used as an example, as shown in Figure 10.1. The goal of each dart throw is to hit the center of the board. Common-cause variation is subject to chance with undiscoverable random causes. This is illustrated in Figure 10.1a by a random distribution of hits clustered around the center of the dartboard. Special-cause variation, however, does not fit into this predictable random variation, as shown in Figure 10.1b. Special-cause variation can be assigned directly to some event or phenomenon. Shewhart believed that these assignable causes could be discovered and removed with an economic benefit. Once sources

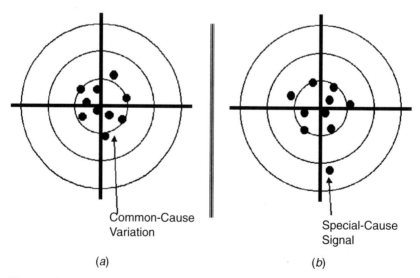

Figure 10.1 Dart board comparison of (a) common-cause variation and (b) special-cause variation.

of special-cause variation are eliminated, improvements can be made to the system to reduce common-cause variation. This is illustrated in Figure 10.2. The random cluster of hits around the center of the dartboard (Figure 10.2a) may be made tighter (Figure 10.2b) by stepping closer to the dartboard.

With this understanding, Shewhart worked to develop a method that would differentiate between the two types of variation. He believed that only through the use of statistics could one obtain an accurate picture of varying physical phenomena. Shewhart toyed with several statistical tools and found success when combining probability analysis with sampling.

On May 16, 1924, Shewhart generated the first basic control chart that used statistically generated graphs to display variations in the quality of manufactured parts. Control charts offered workers the ability to track the performance of a process over time and presented the data in a manner that could be understood at a glance. If the process exhibited common-cause variation, nothing was done. If special-cause variation was identified on the control chart, workers would take action. The result at Western Electric was lower scrap rates and reduced inspection, the economic benefits of which are clear.

In 1925, Shewhart joined Bell Laboratories and continued to apply and refine the control chart. In 1931, Shewhart published the *Economic Control of Quality of Manufactured Products*.[2] This

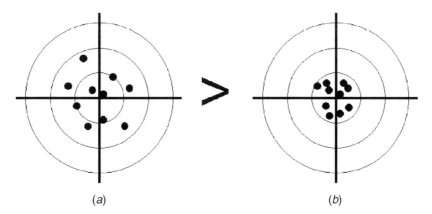

Figure 10.2 Dart board comparison showing a reduction in common-cause variation from (a) to (b).

detailed book introduced the rudimentary concepts of what would become known as statistical process control.

10.2 SPC CHARACTERISTIC TYPES

Process variation can be quantified directly by measuring a process characteristic or indirectly by measuring a product characteristic. With variation quantified, statistical methods can be applied to determine the common-cause and special-cause variations within a process.

Statistical monitoring of product characteristics (product SPC) is the most common application of statistical methods in manufacturing today and can offer insight into the behavior of a process indirectly. Product SPC operates under the premise that a process is in statistical control when a particular product characteristic created by the process is in statistical control. Even though the customers of manufacturers often require that select product characteristics be monitored using statistical methods, data are often collected solely to meet the requirement. Statistical process control in this case becomes statistical process recording.

Less common is the statistical control of process characteristics. Process SPC offers direct insight into the behavior of a process. Process characteristics in some cases are static, such as liquid metal temperature in a holding furnace. However, many processes characteristics are dynamic with a predictable repeating cycle once the process has reached its steady state. When using traditional process SPC methods to control these dynamic characteristics, a measurement is typically taken at an event that repeats during each batch or cycle. An example of such a characteristic is the maximum plunger pressure reached during metal intensification. Although this information may prove useful, it offers little understanding of the entire process.

Process data curves, that is, plots of dynamic process characteristics as a function of time over one cycle, can be examined in order to increase the understanding of a process. Applying statistical methods to entire process data curves, although not feasible a decade ago, can be performed using current computer technology. Process SPC through the analysis of process data curves has been documented in the literature.[3,4]

10.3 SPC APPLIED TO DYNAMIC PROCESS CHARACTERISTICS

The most common SPC method used in the industry is the $\bar{x}$ chart. These charts are typically used to determine if a process is exhibiting common-cause variation or special-cause variation. Figure 10.3 is an example of a typical $\bar{x}$ chart. These charts are prepared by

1. sampling subgroups of four or five individuals,
2. measuring a characteristic on each individual,
3. calculating and plotting $\bar{x}$ (the average of each subgroup),
4. repeating the above three steps for 25 or more samples,
5. calculating $\bar{\bar{x}}$ (the average of each $\bar{x}$) and plotting $\bar{\bar{x}}$ as a line, and
6. calculating the upper and lower control limits at three standard deviations from $\bar{\bar{x}}$ and plotting the control limits as lines.[5]

This SPC method requires that subgroups of multiple samples be formed from data collected under the same conditions and from the same batch or lot.

Through the interpretation of $\bar{x}$ charts, special-cause variation can be identified by using the following guidelines:

1. a point falls outside the control limits,
2. 9 consecutive points occur within one standard deviation,
3. 6 points in a row show a continuous increase or decrease,

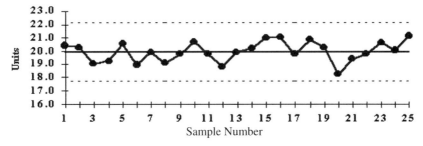

Figure 10.3 Example $\bar{x}$ chart commonly used for SPC.

4. 15 points occur within one standard deviation of the centerline ($\bar{\bar{x}}$), or
5. data follow a cyclic or periodic pattern.

When any of the guidelines above are met, the process exhibits special-cause variation and is termed "out of control."[5]

Traditional $\bar{x}$ charts utilize subgroups of four or five individuals collected under similar conditions. Process data curves, however, stand on their own. Each curve is generated under unique conditions. As such, a statistical method must be used that charts individuals or subgroups of one.

X charts (very similar to traditional $\bar{x}$ charts) are plots of individual data points and are acceptable for use when only one data point can be obtained for a given condition.[6] The control limits for an X chart are defined as two standard deviations from the average. This method of process control can be applied to process data curves for dynamic characteristics.

Statistical analysis of an entire process data curve can be performed by applying finite element methods. By employing the finite element method known as discretization, a discrete model can be generated with a finite number of elements, or nodes, that approximates a corresponding continuous (analog) model.[7] In this case the continuous model is the process data curve. A control chart for a process data curve can be developed by performing the following steps:

1. Collect process data.
2. Divide data into equivalent curves (one curve for each cycle).
3. Create a discrete a model of each curve.
4. Calculate the average for equivalent elements using the discrete models.
5. Create a discrete process average curve using data from the previous step.
6. Calculate the standard deviation for equivalent elements using the discrete models.
7. Create discrete process control limit curves two standard deviations above and below the process average curve using data from the previous step.

Once these steps are completed, a chart can be plotted of the discrete process average model and the discrete control limit models. Process data curves can be plotted with the control limit curves. If any portion of a process data curve falls outside the control limit curves, special-cause variation has occurred during the cycle. Once variation is distinguished as common-cause or special-cause, improvement efforts can be focused at the process to eliminate the assignable special-causes or reduce common-causes.

10.4 DIE SURFACE TEMPERATURE CASE STUDY

To facilitate discussion, the method presented will be applied to a specific example related to high integrity die casting processes. Die surface temperature is a dynamic process characteristic and can be used to illustrate the statistical method presented in this chapter. An example of a process data curve for die surface temperature over one cycle can be found in Figure 10.4. Time 0.0 is defined as when metal enters the die. The metal in the die quickly begins to cool along with the die surface. When the metal solidifies, the heat of fusion liberated slows the cooling of the die surface.[8] The now solid metal and the surface of the die continue to cool. When the die is opened and the solid component is ejected, the die surface will continue to cool until an equilibrium

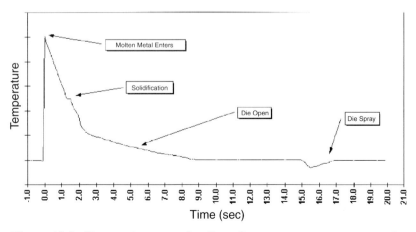

Figure 10.4 Process data curve for die surface temperature over one cycle.

temperature is reached with the bulk mass of the die. Typically, lubricants are sprayed onto the surface of the die between metal injections. Although this may cool the surface of the die quickly, once the spray stops, the surface temperature will rise back to the temperature of the bulk die mass. All the events described above can be associated with features illustrated in the process data curve.

Often die surface temperature, if measured at all, is controlled by monitoring the maximum die temperature for each cycle. Although this information may prove useful, it offers little understanding of the process in comparison to the entire process data curve. Through the examination of die temperature as a function of time, the understanding of the process is increased.

Data for die surface temperature can be obtained by instrumenting a die with a thermocouple. During processing, analog data from the thermocouple can be recorded using a computerized data acquisition system. Such systems store the data in a digital format. Storing the data in this manner in reality creates a discrete model of the original continuous signal from the thermocouple.

The time of each cycle will not be identical. To apply the finite element method, each process curve should be described with an equal number of elements. Thirty process data curves for die surface temperature as shown in Figure 10.5. After collecting the data, the average cycle time was calculated to be 20 s. To help correlate the elements to time, each process data curve has been

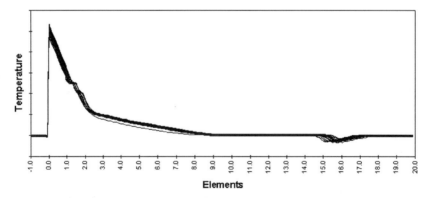

Figure 10.5 Thirty discrete process data curves for die surface temperature over one cycle (equal number of elements for each cycle).

10.4 DIE SURFACE TEMPERATURE CASE STUDY 153

broken into 200 discrete elements with each element equivalent to 0.1 on the x axis.

Table 10.1 is composed of the discrete data from each of the 30 process data curves at element 0.0, when molten metal enters the die. The footnote to the table contains the average of these elements, the standard deviation, and the upper and lower control limits defined as two standard deviations above and below the average. Such calculations are to be performed for each set of 30 equivalent elements.

Figure 10.6 is a plot of the process average curve, upper control limit curve, and lower control limit curve for each element. Note that the common-cause variation described by the control limit curves is not constant. The most common-cause variation can be found during die spray, which is enlarged in Figure 10.7. This shows that more variation in die surface temperature occurs during the die spray than during any other portion of the cycle.

The process data curves should fall within the upper and lower control limit curves. Figure 10.8 is a plot of the process control limit curves and a process data curve exhibiting special-cause variation. The die surface temperature for the data curve plotted falls below the lower control limit curve after solidification. Statisti-

TABLE 10.1 Calculation of Upper and Lower Control Limits for Element 0.0

1	1000	16	1022
2	1011	17	1013
3	1023	18	1025
4	1030	19	1012
5	1014	20	1008
6	999	21	1024
7	983	22	997
8	976	23	973
9	972	24	982
10	970	25	994
11	982	26	1006
12	989	27	1001
13	1003	28	992
14	1018	29	987
15	1024	30	982

Note: average, 1000.40; standard deviation, 18.00; upper control limit, 1036.41; lower control limit, 964.39. Metal enters the die.

154 APPLYING STATISTICAL PROCESS CONTROL

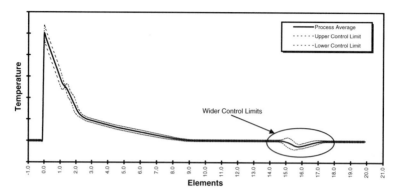

Figure 10.6 Process average curve, upper control limit curve, and lower control limit curve for die surface temperature over one cycle.

cally, the cycle is "out of control" and the special-cause, which induced the variation, should be corrected.

10.5 APPLYING SPC TO HIGH INTEGRITY DIE CASTING PROCESSES

The figures presented in this chapter illustrate the benefits of monitoring dynamic process data curves and applying statistical methods for controlling high integrity die casting processes. Although this discussion has been limited to one example, several other

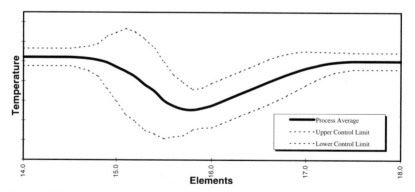

Figure 10.7 Process average curve, upper control limit curve, and lower control limit curve for die surface temperature during die lubricant spray.

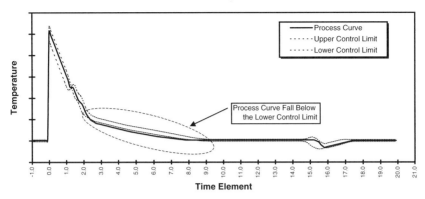

Figure 10.8 Process data curve exhibiting an "out-of-control" condition (cycle exhibits special-cause variation).

dynamic characteristics related to high integrity die casting processes can be analyzed using this technique:

1. plunger velocity,
2. shot sleeve temperature,
3. die cavity pressure,
4. hydraulic pressures, and
5. billet heating for semi-solid metalworking.

These process data curves can be tracked and examined statistically to focus improvement efforts.

REFERENCES

1. Vinarcik, E., "Walter Shewhart," *Technology Century,* October, 1999, p. 46.
2. Shewhart, W., *Economic Control of Quality of Manufactured Products,* D. Van Norstrand, New York, NY, 1931.
3. Robbins, T., "Signature-Based Process Control & SPC Trending Evaluate Press Performance," *MetalForming,* January 1995, pp. 44–50.
4. Vinarcik, E., "Finite Element Analysis of Process Data Curves for Statistical Process Control," SAE Paper Number 970081, Society of Automotive Engineers, Warrendale, PA, 1997.
5. Ishikawa, K., *Guide to Quality Control,* Kraus International Publications, White Plains, NY, 1986.

6. Juran, J., and F.Gryna, *Juran's Quality Handbook,* McGraw-Hill, New York, NY, 1988, p. 24.18.
7. Bickford, W., *A First Course in Finite Element Method,* Irwin, Boston, MA, 1990, p. 7.
8. Taylor, H., M. Flemings, and J. Wulff, *Foundry Engineering,* Wiley, New York, 1959, p. 75.

11
DEFECTS IN HIGH PRESSURE CASTING PROCESSES

11.1 INTRODUCTION

Casting processes are inherently complex due to the phase transformation from liquid to solid metal, which creates all geometric features as well as material properties. With such an intricate process, many potential defects may result. Potential defects related to high integrity die casting are presented in this chapter. Defects encountered in conventional die casting will be introduced but will not be discussed in detail, as several texts are available that examine these issues. Defects unique to high integrity die casting processes will be discussed in greater detail.

11.2 CONVENTIONAL DIE CASTING DEFECTS

Many potential defects commonly found in conventional die casting can also occur in high integrity die casting. Often these defects are avoidable if basic guidelines are followed related to component design, equipment selection, die design, process design, raw material quality, process control, die maintenance, equipment upkeep, and material handling. Conventional die casting defects can be divided into three distinct categories: surface defects, internal defects, and dimensional defects. Each category will be discussed separately below.

11.2.1 Surface Defects

Defects visually identified on the exterior of the component are grouped into the category of surface defects. In some cases, a surface defect does not render the component unfit for use unless aesthetics are a necessary requirement. Discarding a component because it "looks bad" when aesthetics is not a requirement does nothing but create waste.

Cold shuts (also known as cold laps or knit lines) are imperfections visible on the surface of the casting due to unsatisfactory fusion of partially solidified metal. During filling of the die, the convergence of two nearly solidified fill fronts may not knit properly, resulting in this defect. Although cold shuts are characterized by their appearance on the surface of a component, this defect often extends into the bulk metal, creating a weak spot.

Surface contamination commonly identified by discoloration may occur in die casting processes. Often this staining is caused by manufacturing lubricants. Altering the type of lubricants often resolves this issue.

Cracks often occur in die cast components. In some cases, cracks may be caused by cold shuts or residual stresses that form in the component during solidification and cooling. More often than not, cracks are caused by poor material handling techniques. Due to the high production rates, components are ejected from the die near their solidification temperature. While in this fragile state, care must be taken in handling the components. Cracks may occur due to uneven ejection from the die or due to an impact if they are dropped immediately after ejection.

Drags are the result of mechanical interference between the component and the die cavity during ejection. Severe drags may also cause distortion or cracking of the component as well. Proper die design and maintenance are a must to avoid this issue.

Flash is the undesired formation of thin metal sections. Flash most often forms along the parting line and between other independent die components. In most cases, flash is the result of high metal temperatures, high metal intensification pressures during solidification, dimensional variations in the die, and general die wear. Regular die maintenance may eliminate or minimize the occurrence of flash. However, flash is a chronic problem in all die casting technologies. Secondary operations such as trimming or shot blasting are typically used to remove this material.

11.2 CONVENTIONAL DIE CASTING DEFECTS

Laminations are a type of cold shut that occurs at the surface of the component. Due to the turbulent and complex flow patterns within a die during metal injection, a portion of the die may receive a small quantity of metal that freezes quickly to the die surface, forming a thin layer of skin. As the die continues to fill, a bond does not form between this thin layer of solidified skin and the subsequent metal filling the cavity. The result is a thin, partially attached lamination.

A short shot is the incomplete filling of the die cavity caused by an undersized volume of metal being metered into the metal injection system. Such defects are easy to detect by examining the ejected component(s) and the runner system. Without enough metal in the injection system, the biscuit at the start of the runner system may be absent or is very small. In most cases, the component is not fully formed.

Sinks are surface defects caused by localized solidification shrinkage beneath the surface of the casting. Although this defect is common in conventional and vacuum die casting, this problem may be alleviated by using squeeze casting and semi-solid metalworking.

Repetitive thermal cycling of dies from the injection of metal followed by the application of lubricants results in fine cracking of the die face. This phenomenon is referred to as heat checking. Once these cracks are present, metal will fill these fine cracks creating veins or fins on the manufactured component. Although die maintenance may delay the onset of heat checks, die cavities must be replaced to correct this issue. If aesthetics is not a concern, veins from heat checking may not render a component unfit for use.

Lubricants are typically applied to the surface of a die to avoid interaction between the die surface and the metal injected into the die. However, interactions may still occur resulting in a metallurgical bond between the die and the injected metal. This phenomenon is known as soldering and occurs most often when injecting aluminum into ferrous dies.

11.2.2 Internal Defects

Several defects can occur below the surface of a component resulting in less than ideal mechanical properties. Such defects are not visible to the manufacturer, making their detection difficult.

Contamination occurs when unwanted debris is mixed with the metal during injection. Common sources of contamination include degrading refractories, manufacturing lubricants, fractured equipment, and unclean remelted scrap. In most cases, contamination can be controlled by general good housekeeping practices, preventative maintenance, and proper metal melting methods.

Several techniques may be utilized to clean and prepare metal for injection into the die, including the use of fluxes. Excessive amounts of flux may contaminate the metal. Components manufactured with flux-contaminated metal have less than ideal mechanical properties and higher susceptibility to corrosion.

When heating metal for injection into the die, surface oxidation may occur. If the metal is not cleaned, the oxide may scatter throughout the component during injection, forming inclusions. Since most metal oxides are abrasive, inclusions often cause machining problems and excessive wear on cutting tools.

Porosity is a potential defect commonly found in conventional die castings. In many cases, the functionality of a component is not affected. However, porosity is a serious problem in pressure vessels and structural members. Porosity may be attributed to two main sources: solidification shrinkage and gas entrapment.

Most alloys have a higher density in the solid state as compared to the liquid state. This results in shrinkage during solidification. Centerline porosity cavities can occur in alloys that freeze over a narrow temperature range, as in eutectic alloys and pure metals. Interdendritic porosity can occur in alloys that freeze over a wide temperature range. Inadequate feeding of metal to the die cavity during solidification will result in porosity due to shrinkage.

Entrapped gas can originate from several sources and cause porosity. Air can become physically entrapped in the metal during injection. Gases soluble in the liquid alloy may exceed their solubility limit during solidification and evolve as a gas, resulting in porosity. Reactions can occur between the metal and slags, producing a gas, which causes porosity in the final component. Decomposition of lubricants and chemicals used during manufacture can result in gas formation and entrapment in the metal. In most cases, porosity from entrapped gas is caused by multiple sources, making its elimination difficult.

When dealing with liquid alloys, sludging may occur if intermetallic compounds are allowed to precipitate in the metal. In some extreme cases, sludge formation may alter the chemical composition of the alloy. If sludge is mixed with the metal and injected into the die cavity, the final component may have hard spots. These hard spots often cause problems during machining, including excessive wear on cutting tools.

11.2.3 Dimensional Defects

Some defects are related to the geometric dimensions of the component. Even though a die may produce acceptable parts when initially beginning production, several defects may arise over time.

Repetitive thermal cycling can result in catastrophic failure of the die. Such failures occur most often to cores and along die features with sharp corners. Thermal management within cores is difficult. Often small cores cannot be cooled with water lines and must be cooled exclusively with die lubricant sprays. Combined with soldering, cores may fail. When this occurs, the core is retained in a finished component during ejection from the die. All subsequent components are produced without the cored feature.

Repetitive flow of metal across the surface of the die results in erosion. Depending on the metal alloy system, erosion can be both mechanical and chemical in nature. Although component design, die design, and die maintenance may delay the onset of this defect type, high volume production die will eventually show signs of erosion. Component geometries must be monitored to assure compliance to specifications.

Components manufactured using die casting technologies may exhibit warpage. This phenomenon is caused by asymmetric geometric features that contract at different rates during cooling. Warpage can be corrected by part design. If this is not possible, attempts can be made to compensate for the warpage during manufacture.

11.3 DEFECTS OCCURRING DURING SECONDARY PROCESSING

Not all defects arise immediately after manufacturing a component. Defects can emerge during secondary operations. In some cases, a component must be heat treated or welded. In such cases, blisters may form on or near the surface of a component. When metal is injected into the die, entrapped gases are often compressed into very small bubbles. When the temperature of the component is elevated during welding or heat treatment, compressed gases expand, forming blisters. For this reason, most conventional high pressure die castings do not undergo secondary thermal processes.

As discussed earlier, cracks often occur after a component is manufactured. In many cases this includes secondary processing. Trimming is performed to remove flash as well as the runner system from a component. Trimming may cause cracks if the fixturing does not support the component evenly or if trim punches are oversized. Machining operations often damage and crack components. Hydraulic clamps can easily overstress a component, causing cracks. Moreover, vibrations caused by cutting tool chatter during machining may induce cracks.

11.4 DEFECTS UNIQUE TO SQUEEZE CASTING AND SEMI-SOLID METALWORKING

When commercializing squeeze casting and semi-solid metalworking processes, component producers looked to conventional die casting to identify potential defects and control component quality. Several defects were expected, including cold shuts, cold flows, flash, drags, warping, and gas entrapment, to name a few.[1] Efforts were taken to avoid these defects by addressing processing methods, die design, and product design. Defect types, however, have surfaced unique to squeeze casting and semi-solid metalworking processes.

Although squeeze casting and semi-solid metalworking have proven to be successful for many commercial applications, component producers have been reluctant to report defects for fear of giving these emerging processes a bad reputation. Nonetheless, these defects are easily classified and must be understood to avoid future problems. Contaminant veins and phase separation are the defect types presented in this discussion.

11.4.1 Contaminant Veins

Contaminant veins are unique defects that result from the planar filling phenomenon of squeeze casting and semi-solid metalworking processes. To understand this defect, one may examine the method of its formation. When filling the die, the metal front remains relatively intact (Figure 11.1a) and picks up contaminants such as die lubricants, die steel corrosion products, and other impurities, as illustrated in Figure 11.1b. This contamination is exacerbated by oxidation at the metal front. As filling of the die is

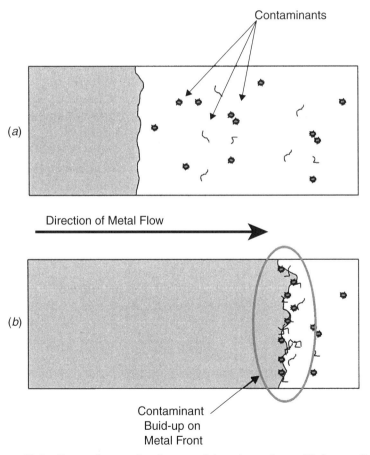

Figure 11.1 Contaminant veins form as (a) a clean planar fill front collects contaminants and (b) the metal progresses through the die cavity.

nearly complete, this contaminated metal front typically converges with another contaminated metal front. The result is a contaminant vein composed of metal oxides and other impurities trapped within the casting. Figure 11.2 is a graphical illustration showing the locations prone to this defect. Most often contaminant veins are located at the last location of fill within a die. In some cases, contaminant veins are near or attached to cores opposite the direction of metal flow. This results when the fill front is split and converges on the far side of the core.

Contaminant veins should not be confused with cold shuts or cold laps as temperature is not the driving force in the creation of the defect. In addition, contaminant veins are not equivalent to the inclusion-type defects often observed in castings as contaminant veins occur with consistency for a given component and gating geometry combination.

In many cases, the presence of a contaminant vein does not affect the functionality of the product. Contaminant veins often go unnoticed, trapped beneath the surface of the product. Secondary processes such as machining or trimming, however, can open a path to the defect. In such cases, leak tightness can be compromised. The presence of a contaminant vein in a structural member can be a major problem depending on its location. The contaminant vein has inferior mechanical properties in comparison to the bulk material. Moreover, the vein can act as a stress concentrator.

Several actions related to processing, die design, and component design can be taken to resolve issues arising from contaminant veins.

Contamination in the die should be minimized. Lubricants should be used sparingly. Die sprays should not be used to cool

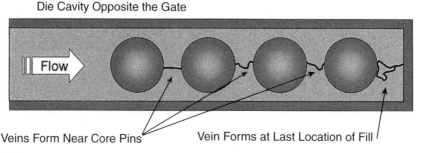

Figure 11.2 Characteristic features prone to contaminant vein formation.

the die but should only be used to apply a lubricant to ease ejection of the product after forming. Dies should be thermally balanced using water lines to achieve die cooling.

Effort should also be taken not to introduce metal oxides into the die with the metal being formed. Oxidation products should be removed from the surface of molten metal prior to ladling in the squeeze casting process. In some cases, a protective layer of an inert cover gas such SF_6 can be maintained over the molten metal to minimize oxidation. Specific to the indirect semi-solid processes, billets may be scrapped to remove built-up surface oxides that have formed during handling and heating. For direct processes, care should be taken to minimize oxide formation during heating by avoiding atmospheric contact. Inert atmospheres can be used as well.

Gating design can significantly affect contaminant vein formation. The distance the metal front must flow to fill the runner system and die cavities must be minimized. The longer the fill path, the greater the opportunity to collect contaminants on the metal fill front. The metal front in many cases can be manipulated to stretch the fill front over a great distance, such as with the use of a fan gate. Stretching the metal fill front disperses the contaminants over a larger area and avoids the formation of a concentrated contaminant vein.

In many cases, contaminant veins cannot be avoided. However, the location of the vein can be controlled by gating. In such cases, the contaminant vein can be placed in an area of the product that will not be machined or, as in the cases of structural members, does not see high stresses. Overflows can also be utilized to capture contaminant veins. The overflows can then be removed from the finished product, yielding a component absent of the defect.

Every processing defect can be exacerbated or minimized through product design. Product designers and process engineers must work together as early as possible to maximize manufacturability while meeting functional requirements. Geometric features should be incorporated into the design to assist in gating and minimizing metal flow distance during forming.

11.4.2 Phase Separation

Unique to semi-solid metalworking processes is a defect known as phase separation. The metal injected into the die cavity during

166 DEFECTS IN HIGH PRESSURE CASTING PROCESSES

semi-solid processing is partially solid and partially liquid. This two-phase mixture does not necessarily remain homogeneous. The liquid phase flows easily and in some cases will leave its solid counterpart behind, resulting in phase separation.

Cases of phase separation have been observed when the metal fill front must travel a significant distance within the die cavity while flowing around multiple cores. Figure 11.3 is a graphical illustration of this phenomenon. The cores choke the passage of the solid phase. The resulting product has nonuniform material properties as the microstructure of the metal near the gate is highly spheroidal while the microstructure of the metal at the last location to fill is dendritic. Solidification shrinkage also becomes a problem in the dendritic region.

In many cases, the phase separation does not affect the functionality of the products produced using semi-solid metalworking. The dendritic structure may go unnoticed. However, secondary processes such as machining or trimming can open a pathway to

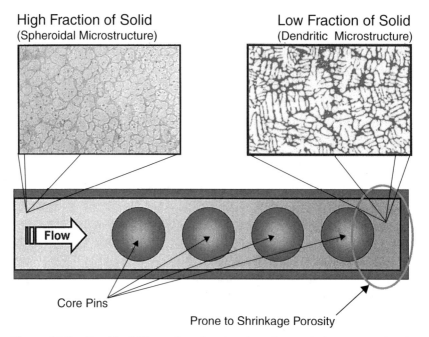

Figure 11.3 Graphical illustration showing the effects of phase separation in semi-solid metalworking.

solidification shrinkage associated with a dendritic region. In pressure vessels, this phenomenon can cause significant problems.

Variation in the mechanical properties caused by phase separation can be a serious issue in structural members. The dendritic region of the component has inferior mechanical properties in comparison to the bulk material. Moreover, porosity caused by solidification shrinkage can act as a stress concentrator, increasing the chances of failure.

Actions can be taken related to die design to minimize phase separation during semi-solid metalworking. Gating can significantly affect phase separation induced during fill. The distance the metal front must flow to fill the runner system and die cavities must be minimized. Each time the metal front must change direction, the potential for phase separation increases. Choke points such as cores should be minimized. Necessary choke points should be fed with metal on both sides, such as with the use of a fan gate.

Phase separation may be unavoidable for some applications. Overflows can be utilized to capture the dendritic region of the component. Unfortunately, the use of overflows reduces processing yield. The location of the dendritic region, however, can be manipulated by varying the location of the gate. In some cases, the dendritic region can be selectively located in a noncritical location within the product.

11.5 PREDICTING DEFECTS

Both product designers and component producers must be aware of the potential defects that may occur in components manufactured using high integrity die casting processes. Many potential defects are well understood. Other defect types are unique to squeeze casting and semi-solid metalworking. With a clear understanding of the mechanisms that induce defects, efforts may be made prior to product launch to minimize their occurrence.

Flow modeling software has developed significantly over the past decade, and many software packages include data for analyzing vacuum die casting, squeeze casting, and semi-solid metalworking processes. These analytical tools should be utilized to predict the formation of contaminant veins, ideally while the designs of both the product and the die can be modified. The prediction of phase separation using computer flow modeling is not yet possible. The currently available computational fluid dynamic software defines the semi-solid metal as a high viscosity fluid rather than as a true two-phase mixture. A method of modeling two-phase flow was proposed in 1997 and efforts are currently underway to develop the proposal into a viable computer model.[2]

REFERENCES

1. Keeney, M., J. Courtois, R. Evans, G. Farrior, C. Kyonka, A. Koch, K. Young, "Semisolid Metal Casting and Forging," in Stefanescu, D. (editor), *Metals Handbook,* 9th ed., vol. 15, *Casting,* ASM International, Materials Park, OH, 1988, p. 327.
2. Alexandrou, A., G. Burgos, and V. Entov, "Semisolid Metal Processing: A New Paradigm in Automotive Part Design," SAE Paper Number 2000-01-0676, Society of Automotive Engineers, Warrendale, PA, 2000.

VISIONS OF THE FUTURE

12
FUTURE DEVELOPMENTS IN HIGH INTEGRITY DIE CASTING

12.1 CONTINUAL DEVELOPMENT

Driven by economics, efforts are underway to further improve several aspects of the high integrity die casting processes discussed in this text. Presented in this chapter are numerous technologies related to both processing and materials currently being researched and developed.

12.2 NEW HIGH INTEGRITY DIE CASTING PROCESS VARIANTS

Among the high integrity die casting processes discussed in this book, semi-solid metalworking is the most costly. For this reason, it is the subject of intense research. Thixomolding® is among the most economical semi-solid metalworking variants since the semi-solid metal mixture is produced on demand. This process, however, is currently limited to use with magnesium casting alloys. Research is being conducted to find a suitable barrel-and-screw material for use with aluminum casting alloys. Several investigations have failed to identify a suitable material, including a research project sponsored by the U.S. Department of Energy. Regardless of the setbacks, research in this area continues.

Other direct semi-solid metalworking processes are under development for use with aluminum casting alloys. Research has focused on developing production equipment that can convert liquid metal into the semi-solid slurry on demand. Ube Industries has introduced a four-step version of the semi-solid processing method[1,2] (Figure 12.1). Initially, metal is ladled into a ceramic crucible just above its melting point. The liquid metal is cooled to a target temperature using controlled air knives. Metal temperature is then stabilized and the liquid metal is stirred by induction heating. The resulting semi-solid slurry is poured into a vertical shot sleeve by turning the cup upside down (to isolate oxides at the plunger tip) and injected into the die.

12.3 REFINEMENTS OF MAGNESIUM ALLOYS

Magnesium alloys have many limitations that prevent its use in many applications. Several issues must be addressed to expand the capabilities of magnesium alloys related to corrosion resistance and creep.

Magnesium is very susceptible to galvanic corrosion. Fastening methods must be developed and designed that inhibit galvanic corrosion when high integrity magnesium die castings are joined with components of a dissimilar metal. Barrier coatings are being utilized to solve this problem in some applications.[3] However, minor scratches or porosity in currently available barrier coatings

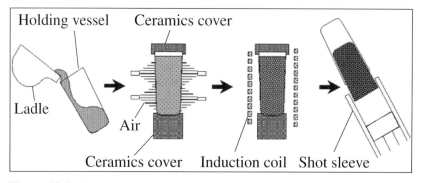

Figure 12.1 Four-step process for producing a semi-solid slurry on demand. (Courtesy of UBE Machinery, Inc.)

result in accelerated corrosion at the defective location. Research is being conducted to address this problem.

Current commercial magnesium alloys are limited to near-room-temperature applications due to decreased creep resistance as the temperature is elevated. Some magnesium alloys containing rare earth elements are available that can operate at temperatures as high as 150°C with a significant cost penalty. Past research has shown that additions of aluminum and alkaline rare earth elements (barium, strontium, and calcium) in magnesium form fine precipitates of $Al_{11}E_3$ at grain boundaries.[4,5] These precipitates inhibit creep at elevated temperatures. Of the alkaline metals currently researched, calcium is the least costly as well as the least dense with the lowest total "cost per atom" of the possible choices. With this knowledge, several magnesium alloys containing aluminum and calcium are under development.

12.4 EMERGING ALLOYS FOR USE WITH HIGH INTEGRITY DIE CASTING PROCESSES

Aluminum and magnesium are the dominant metals utilized in high integrity die casting. However, efforts are underway to harness the economic benefits of high integrity die casting processes with nontraditional die casting alloys.

Vacuum die casting is being utilized to produce components in titanium, β-titanium, titanium aluminide, nickel-based alloys, amorphous metallic glasses, and stainless steel.[6,7] The die casting dies utilized in with these alloys are fabricated from high temperature refractory metals such as tungsten or molybdenum. Products manufactured include airfoils, fan blades, structural hardware, golf club heads, and automotive valves. Current production quantities are low with limited shot sizes.

12.5 METAL MATRIX COMPOSITES FOR USE WITH HIGH INTEGRITY DIE CASTING PROCESSES

Within the materials community, composite materials are a subject of considerable research. As a result of these efforts, discontinuously reinforced metal matrix composites are under development for commercial use with high integrity die casting processes. Discontinuous reinforced casting alloys are a class of metal matrix composites in which a metal alloy matrix is reinforced with ceramic particles or whiskers. Figure 12.2 is an illustration of a component produced using such a composite with the die casting process. Discontinuously reinforced composite materials exhibit improved characteristics when compared to traditional casting alloys, including

reduced structural weight,
increased tensile modulus,
improved yield strength,
increased ultimate tensile strength,
improved fatigue limit,
improved dynamic response, and
enhanced wear resistance.

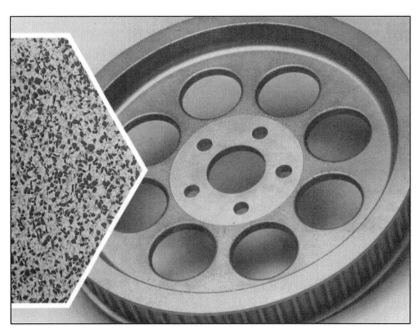

Figure 12.2 Motor cycle sprocket die cast using an SiC particulate reinforced aluminum matrix composite. (Courtesy of Alcan, Inc.)

Several aluminum and magnesium metal matrix composites are commercially available utilizing several ceramic reinforcing materials. Reinforcing materials used with magnesium have been limited to SiC.[8] However, reinforcing materials commonly used with aluminum alloys include SiC, Al_2O_3, B_4C, and flyash.[9-11] Of these reinforcing materials, flyash is the most economically favorable filler as it is a waste product from coal-burning power plants. Flyash has a raw material cost ranging from $15 to $30 per ton.

Of the high integrity die casting processes presented in this text, semi-solid metalworking is best suited for use with metal matrix composites. Maintaining homogeneity of the composite is difficult when the composite material is molten due to differences in densities. The thixotropic properties of the semi-solid metal slurry and the stirring inherent in semi-solid metal preparation create ideal conditions for maintaining composite homogeneity.

12.6 REDUCING TOOLING LEAD TIMES

Speed to market is a necessity in today's competitive economy. Months, and sometimes weeks, can drastically change a company's competitive edge and profitability. A major quandary to any high integrity die casting process is the time required to build tooling, specifically in the manufacture of the complex casting cavity.

Long tooling lead times frequently force designers to choose other manufacturing strategies. Most manufacturing methods with short tooling lead times are limited in their ability to produce complex geometries. Typically, hastily tooled components are fabricated assemblies composed of numerous subcomponent parts. Although a product may reach the marketplace before competitors, the product may be very costly. In the drive to reduce product lead times, the economic benefits possible with high integrity die castings are often sacrificed.

Over the last decade vast improvements in computer hardware, computer-aided design (CAD), and computer-aided manufacturing (CAM) have made rapid prototyping possible. Often individual components can be manufactured in days or hours, giving today's product engineers and designers the ability to evaluate designs quickly. These same rapid prototyping techniques can be applied to quickly manufacture die casting cavity inserts for use in high integrity die casting processes.[12–14]

As the accuracy of CAD/CAM technology advances, carbon electrodes are being machined with greater precision for use with plunge electrodischarge machining (EDM). Many die-making companies are also using prehardened tool steel inserts manufactured using EDM technology. As carbon electrode technology advances, tool shops may be able to produce die cavities with a single electrode, eliminating the iterative EDM process commonly used today.

Stereo prototyping methods are also being developed for direct production of die cavity tooling. Instead of using liquid resin, laser sintering with metal powder has been used in the development of this technology. Experimentation is underway utilizing several different nontraditional metal powders, including polymer-coated zirconium diboride, bronze–nickel mixtures, and 316 stainless steel.

Casting cavity inserts are also being produced using an indirect form of rapid prototyping. Wax patterns of a casting cavity can be produced using polymer stereo lithography technology. These patterns can then be investment cast, producing die cast tooling inserts in H13 tool steel. However, the investment casting method has several problems. Large metal masses such as a casting cavity inserts are difficult to cast, resulting in porosity and distortion. To overcome these problems, experimentation is underway to test hollow casting cavity inserts investment cast with a uniform thickness.

12.7 LOST-CORE TECHNOLOGIES

A major limitation of high integrity die casting processes is the ability to produce complex internal geometries. Simple coring is regularly performed, but high integrity castings cannot be produced with undercuts. However, most casting processes that do not utilize reusable molds regularly produce complex internal geometries, as illustrated in Figure 12.3. With such casting processes, the core is lost. In the cases of resin-bonded sand, the heat

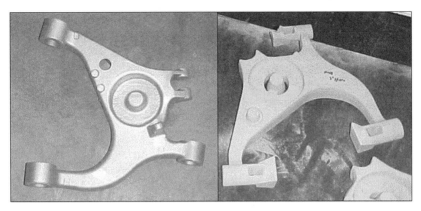

Figure 12.3 Hollow (*a*) aluminum automotive suspension arm and (*b*) resin-bonded sand core. (Courtesy of Teksid.)

of the metal causes the resin to combust and the core can be removed from the casting as loose sand.

Die casting component producers have experimented with lost-core processes for several years. Finding a material is difficult. Several conflicting requirements must be met. The core material must withstand erosion during metal injection. The material must remain dimensionally stable under high pressures. After solidification, the core material must be easily removed. Trials have been conducted using salt core materials and low melting point alloys. Although these materials survive the metal injection, removal of the cores is time consuming and costly.

Resin-bonded sand was abandoned years ago because sand cores could not stand up to the high metal injection velocities encountered in die casting. However, this technology is being re-examined for use with squeeze casting. Since metal velocities are significantly less for squeeze casting, sand cores may survive. Experimentation is currently underway to prove this potential solution.

12.8 CONTROLLED POROSITY

Porosity is a defect commonly found in die cast components. However, if the porosity does not affect fit or function, one cannot call it a defect.

With this philosophy in mind, experimentation is being conducted to produce die cast components with a controlled amount of porosity. This approach, although counterintuitive, is also being studied intensely by the injection molding industry. Benefits of this technology are reduced weight and less material usage. Candidates for this technology are limited to nonstructural applications.

12.9 INNOVATIONS CONTINUE

Numerous other projects are underway related to high integrity die casting processes. Some research is focusing on casting machine development. Other work is underway to extend tooling life. Although some future developments related to high integrity die casting processes can be predicted, one can only speculate as to the state of this art in years to come.

REFERENCES

1. Adachi, M., and S. Sato, "Advanced Rheocasting Process Improved Quality and Competitiveness," SAE Paper Number 2000-01-0677, Society of Automotive Engineers, Warrendale, PA, 2000.
2. Zehe, R. "First Production Machine for Rheocasting," *Light Metal Age,* October 1999, p. 62.
3. Wang, G., K. Stewart, R. Berkmortel, and J. Skar, "Corrosion Prevention for External Magnesium Automotive Components," SAE Paper Number 2001-01-0421, Society of Automotive Engineers, Warrendale, PA, 2001.
4. Powell, B., A. Luo, V. Rezhets, J. Bommarito, and B. Tiwari, "Development of Creep-Resistant Magnesium Alloys for Powertrain Applications: Part 1," SAE Paper Number 2001-01-0422, Society of Automotive Engineers, Warrendale, PA, 2001.
5. Luo, A., M. Balogh, and B. Powell, "Development of Creep-Resistant Magnesium Alloys for Powertrain Applications: Part 2," SAE Paper Number 2001-01-0423, Society of Automotive Engineers, Warrendale, PA, 2001.
6. Larsen, D., and G. Colvin, "Vacuum-Die Casting Titanium for Aerospace and Commercial Components," *Journal of Metals,* June 1999, p. 26.
7. Larsen, D., "Vacuum-Die Casting Yields Quality Parts," *Foundry Management and Technology,* February 1998, p. 32.
8. Wilks, T., "Cost-effective Magnesium MMCs," *Advanced Materials and Processes,* August 1992, p. 27.

9. Rohatgi, P., "Cast Aluminum-Matrix Composites for Automotive Applications," *Journal of Metals,* April 1991, p. 10.
10. Rohatgi, P., "Aluminum-Flyash Composites," *Foundry Management and Technology,* October 1995, p. 32.
11. Rohatgi, P., "Low Cost, Flyash Containing Aluminum-Matrix Composites," *Journal of Metals*, November 1994, p. 55.
12. Gerdts, M., "Moving to the Next Millennium: Rapid Tooling," *Rapid News,* March 1997, p. 40.
13. Hardro, P., and B. Stucker, "Die Cast Tooling from Rapid Prototyping: Part 1," *Time-Compression Technology,* June 1999, p. 45.
14. Hardro, P., and B. Stucker, "Die Cast Tooling from Rapid Prototyping: Part 2," *Time-Compression Technology,* July 1999, p. 36.

STUDY QUESTIONS

Study Question 1

Metal flow during die fill varies greatly for different high integrity die casting processes. Shown in Figure SQ1 are three types of metal flow behavior. Define the type of flow for each and state which high integrity die casting process exhibits that type of metal flow during die fill.

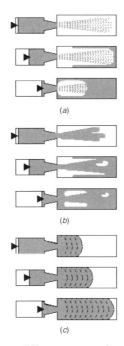

Figure SQ1 Three different types of metal flow behavior.

Solution 1

Atomized, nonplanar, and planar metal flows are illustrated in Figures S1*a*, *b*, and *c*, respectively.

Atomized flow is typical of conventional and vacuum die casting in which liquid metal is traveling at high velocities through a very small gate. Due to the high pressures and velocities, the metal becomes in effect an aerosol, spraying into the die cavity. Filling occurs from the surface of the cavity inward. Typically, the first metal to enter the die strikes the far side of the die cavity and solidifies immediately.

Nonplanar flow is typical of squeeze casting. When squeeze casting, nonplanar fill often results in the die cavity being filled from the outside inward. The metal front enters the die as a single metal stream that begins to fan out. As the metal reaches the far side of the die cavity, gases are entrapped as the fill front doubles over on itself. The metal front continues to travel along the surface of the die filling the cavity from the outside inward.

Planar flow is typically found in semi-solid metalworking. The metal front progresses as a uniform plane filling the cavity completely. Gases in the die are ahead of the metal fill front with minimal entrapment.

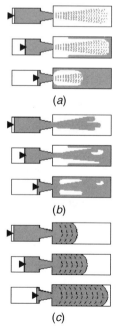

Figure S1 Illustration of (*a*) atomized metal flow, (*b*) nonplanar metal flow, and (*c*) planar metal flow.

Study Question 2

Computer modeling simulations are performed for three different high integrity die casting processes. Shown in Figure SQ2 are the results. To minimize the amount of entrapped gases, a vacuum system will be used to extract air from the die cavity during fill. Specify where the vacuum valve should be located to maximize the benefits of the vacuum for each process.

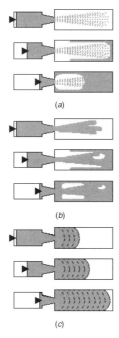

Figure SQ2 Illustration of metal behavior in three processes requiring vacuum valve placement.

184 STUDY QUESTIONS

Solution 2

The optimum location of vacuum valve placement for each case is shown in Figure S2.

Case *a* shows typical fill for conventional and vacuum die casting. Atomized flow results in the metal being sprayed into the die hitting the far side of the cavity. Filling occurs from the surface of the cavity inward. Optimum vacuum valve placement is near the gate at the last portion of the die surface to be covered with metal.

Case *b* illustrates metal fill commonly observed in squeeze casting. Under these conditions, the metal front fans out and reaches the far side of the die cavity before doubling over on itself. The metal front continues to travel along the surface of the die filling the cavity from the outside inward. Optimum vacuum valve placement is near the gate at the last portion of the die surface to be covered with metal.

Case *c* exhibits planar die fill common in semi-solid metalworking. Since the metal front progresses as a uniform plane, the vacuum valve should be located at the point farthest from the gate.

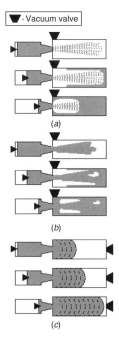

Figure S2 Illustrations showing optimum vacuum valve placement for three metal flow patterns.

Study Question 3

Shown in Figure SQ3 is a manufacturing cell design commonly used with indirect (billet) semi-solid metalworking processes. Number and describe each step in the process.

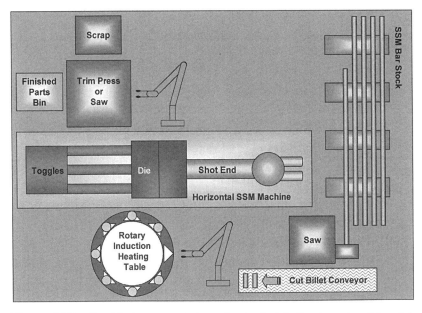

Figure SQ3 Graphical representation of a typical indirect semi-solid metalworking manufacturing cell.

Solution 3

Figure S3 is a numbered illustration of the following processing steps:

1. Semi-solid feedstock in the form of bars is staged for cutting.
2. Semi-solid feedstock is cut into billets of controlled length or volume.
3. Billets are conveyed to a robot.
4. A robot is used to pick a billet from the conveyor for placement onto a rotary table.
5. A rotary table is used to progress billets through a series of induction heating stations.
6. A robot is used to pick heated billets from the rotary table for placement into the shot sleeve of the casting machine.
7. The billet is injected into the die cavity by the plunger of the casting machine.
8. The metal solidifies in the die cavity.
9. The newly cast component with its runner system is picked from the open die by a robot during ejection and placed in an open trim press or saw.
10. The component is trimmed or sawed.
11. The finished component is ejected from the trim press or saw and placed in dunnage.
12. The runner and other scrap are ejected from the trim press or saw and placed in a hopper for recycling.

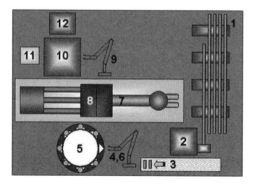

Figure S3 Processing order numbered for a typical indirect semi-solid metalworking manufacturing cell.

Study Question 4

Shown in Figure SQ4 is a manufacturing cell design commonly used with the Thixomolding® process. Number and describe each step in the process.

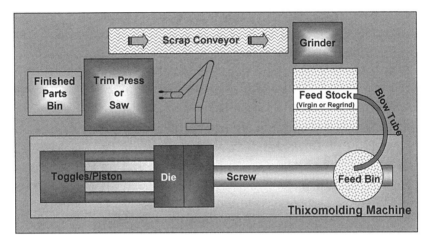

Figure SQ4 Graphical representation of a typical Thixomolding® manufacturing cell.

Solution 4

Figure S4 is a numbered illustration of the following processing steps:

1. Feedstock in the form of metal chips is placed in dunnage.
2. The metal chips are blown from the dunnage.
3. The metal chips are transferred into a feed bin with a controlled atmosphere.
4. The metal chips are metered into the casting machine shot end and heated while a screw stirs the metal.
5. Stirred metal is injected into the die cavity by the screw.
6. The metal solidifies in the die cavity.
7. The newly cast component with its runner system is picked from the open die by a robot during ejection and placed in an open trim press or saw.
8. The component is trimmed or sawed.
9. The finished component is ejected from the trim press or saw and placed in dunnage.
10. The runner and other scrap are ejected from the trim press or saw and placed on a conveyor.
11. Runners and scrap are fed into a grinder.
12. Ground scrap is mixed with virgin feedstock for immediate reuse.

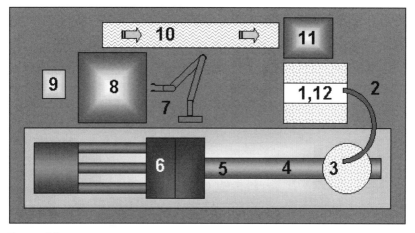

Figure S4 Processing order numbered for a typical Thixomolding® manufacturing cell.

Study Question 5

Both continuously cast and extruded semi-solid metalworking billet material are illustrated in Figure SQ5. Discuss the differences between the billets when they are heated for processing.

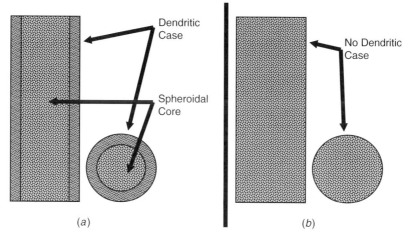

Figure SQ5 Anatomy of (*a*) continuously cast and (*b*) extruded semi-solid metalworking billet.

Solution 5

Continuously cast semi-solid billet material has a dendritic case that forms as a result of the minimal stirring and the high rate of cooling found at the surface of the casting mold. This dendritic case gives the billet structure when heated. The core of the billet will begin to separate when heated due to density differences between the liquid metal phase and the solid phase. Barreling may also occur at the bottom of the billet. When picking up a billet after heating, the core material may drip or flow from the bottom (Figure S5a).

Extruded billet material is homogenous and does not have a dendritic case. When heating this type of billet material, a crucible or other vessel must be used to contain the heated semi-solid slurry (Figure S5b).

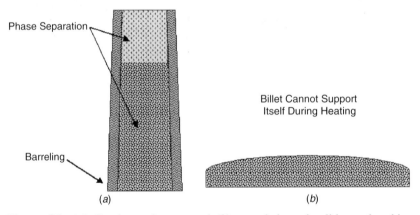

Figure S5 (a) Continuously cast and (b) extruded semi-solid metalworking billet behavior during heating.

Study Question 6

Examine the three aluminum microstructures shown in Figure SQ6. What high integrity die casting processes were used to produce each microstructure? (Fill in the blank boxes in the figure.)

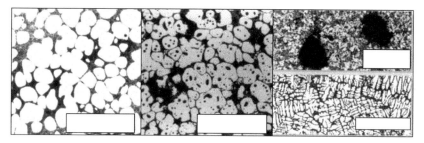

Figure SQ6 Four aluminum microstructures produced using different die casting processes. (Courtesy of UBE Machinery, Inc.)

Solution 6

The processes are noted with their respective micrographs in Figure S6.

An aluminum microstructure produced during a direct semi-solid metalworking has primary aluminum spheroids that make up the solid fraction of the material during manufacture. The surrounding matrix in the microstructure (formerly the liquid portion during manufacture) is composed of fine primary aluminum dendrites and the eutectic phase. This can be distinguished from a microstructure produced during indirect semi-solid metalworking that also has dark microstructural features within the spheroids caused when liquid metal is trapped within the solid fraction. Squeeze casting produces a microstructure with fine but clearly dendritic structure, while conventional die casting produces an extremely fine structure often with entrapped gas or porosity.

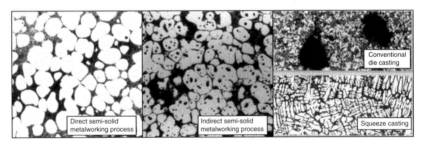

Figure S6 Four aluminum microstructures with die casting process method noted.

Study Question 7

Presented in Figure SQ7 is a die surface temperature curve over time for one cycle of a high integrity casting process. Label each major phase of the cycle. (Fill in the blank boxes.)

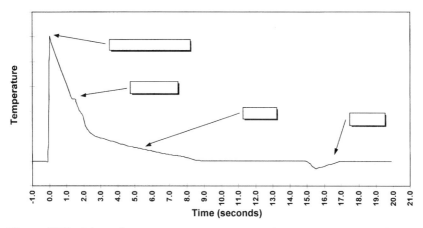

Figure SQ7 Die surface temperature over one casting cycle. (Courtesy of UBE Machinery, Inc.)

Solution 7

At time 0.0, metal enters the die. The metal in the die quickly cools and solidifies. The die is opened and the solid component is ejected while the die surface continues to cool. An equilibrium temperature is reached with the bulk mass of the die. Die lubricant is sprayed onto the surface of the die. Once the spray stops, the surface temperature quickly rises back to the temperature of the bulk die mass. The events described are noted in Figure S7.

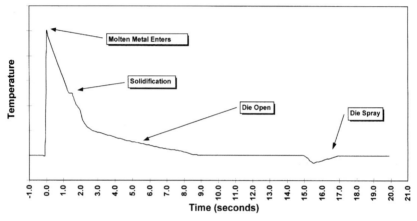

Figure S7 Die surface temperature over one casting cycle with casting phases noted.

Study Question 8

Pictured in Figure SQ8 is a cross section of a weld on the surface of a casting die. Label each major microstructural region. (Fill in the blank boxes.) Define what occurs at the surface in the circled region and state why this phenomenon occurs.

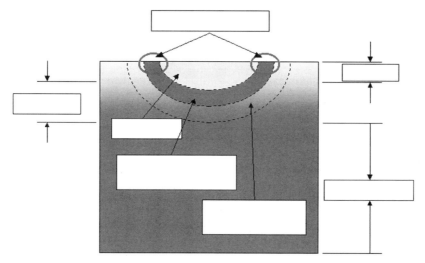

Figure SQ8 Microstructural regions of a welded die face.

Solution 8

Each microstructural region is labeled in Figure S8. The circled regions on the die surface are prone to soldering. Since the metal in the regions circled is annealed, it is soft and easy to erode, unlike the martensitic layer covering most of the die surface. This soft region is prone to soldering.

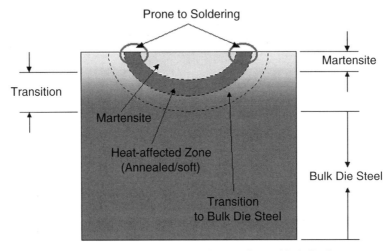

Figure S8 Microstructural make-up of a welded die face.

Study Question 9

Pictured in Figure SQ9 is an air scoop for an agriculture combine produced as a welded fabrication. Apply design-for-assembly principles and discuss how this component can be improved.

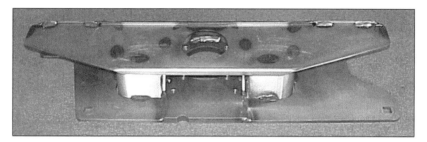

Figure SQ9 Air scoop for an agriculture combine produced as a welded fabrication. (Courtesy of AFS, Inc.)

Solution 9

Design for assembly works to minimize the amount of components required for manufacture. A required component must

1. have movement for function,
2. have a specific material for function, or
3. be serviced.

This analysis may be performed using the flow chart presented in Figure 8.1. Numerous stampings are used in the welded fabrication, as shown in Figure S9.1. Due to the number of individual parts used in the assembly, at least 24 weld points are needed to join the components (Figure S9.2) Warpage caused by thermal distortion during welding resulted in considerable dimensional variation between assemblies.

Pictured in Figure S9.3 is a single piece air scoop. This single component meets the same requirements as the welded fabrication. As no welds are required, the single piece casting has minimal warpage in comparison to the fabrication. Moreover, a 40% reduction in cost was achieved with a significantly reduced investment in tooling.

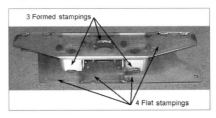

Figure S9.1 Illustration noting the seven individual components of the fabricated air scoop. (Courtesy of AFS, Inc.)

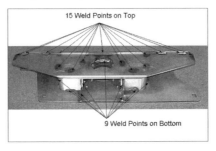

Figure S9.2 Illustration noting 24 weld points on the fabricated air scoop. (Courtesy of AFS, Inc.)

STUDY QUESTIONS 199

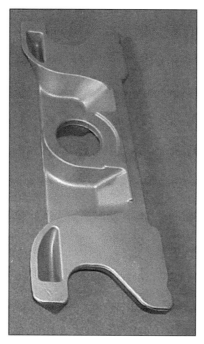

Figure S9.3 Integrated single piece air scoop. (Courtesy of AFS, Inc.)

Study Question 10

What die casting processes should be used to manufacture each of the following products? Choose from conventional die casting, vacuum die casting, squeeze casting, and semi-solid metalworking. Explain your answer for each.

1. Name plate for an automobile
2. Master brake cylinder
3. Fuel rail
4. Cell phone housing
5. Safety critical structural member
6. steering wheel
7. Latch handle
8. Transmission case
9. Automobile lower control arm
10. Automotive steering knuckle

Solution 10

The answer to each question may be conventional die casting. The integrity of conventional die castings vary greatly depending on the skill, knowledge, and discipline of the manufacturer. Not all die casting component producers are equal. As such, purchasing agents and design engineers must be aware of the producer's individual capabilities.

With this point made, the components listed are typically produced using the processes noted:

1. Name plate for an automobile—conventional die casting
2. Master brake cylinder—semi-solid metalworking
3. Fuel rail—semi-solid metalworking
4. Cell phone housing—semi-solid metalworking (in magnesium)
5. Safety critical structural member—squeeze casting or semi-solid metalworking
6. Steering wheel—no process or alloy is dominant for this component
7. Latch handle—conventional die casting
8. Transmission case—vacuum die casting or squeeze casting
9. Automobile lower control arm—semi-solid metalworking
10. Automotive steering knuckle—squeeze casting

APPENDIX A
COMMON NOMENCLATURE RELATED TO HIGH INTEGRITY DIE CASTING PROCESSES

The following section is composed of brief practical definitions for common phrases and terms associated with high integrity die casting processes.

AQL Acceptable quality level, a quality level established on a prearranged system of inspection using samples selected at random.

As cast Cast product without subsequent processing.

Atomized flow Fluid flow in which the fluid is broken into spray of droplets moving in a consistent direction

Biscuit Solidified metal left at the end of the shot sleeve of a cold-chamber die casting machine that is intended to feed the gate and runner system.

Blistering A porous defect on or near the surface of a component resulting from the expansion of compressed gas within the metal.

Charge A given amount of metal introduced into the furnace.

Chill An insert in the die cavity, typically water cooled, used to produce local chilling for increasing the rate of solidification.

Chill vent   An insert in which air is vented from the die cavity and a water chill is used to solidify the metal before it exits the die through the vent.

Cold-chamber die casting   The die cast process variant in which molten metal is ladled into a shot sleeve that injects the molten metal into the die cavity using a plunger.

Cold laps/cold shut   An imperfection in a component due to unsatisfactory fusion of partially solidified metal.

Contaminant vein   A defect caused by the convergence of fouled metal fill fronts.

Core   A metal insert in the die that forms hollow regions in a component.

Cover die   The half of a casting die set that is mounted to the injection side of the die casting machine.

Dendritic   The treelike branch microstructure commonly found in most components.

Dendritic case   The localized dendritic microstructure found along the perimeter of semi-solid feedstock.

Die   A metal form used as a permanent mold for die casting.

Die cavity   The impression in a die filled with molten metal to form a component.

Direct semi-solid metalworking   All semi-solid metalworking process variants in which semi-solid material is prepared just in time for manufacturing a product.

Dowel   A pin of various types used in the parting surface of dies to assure correct registry.

Draft   Taper on the vertical sides of the die cavities that permits the component to be removed without distortion.

Drags   A defect caused by the mechanical interference of the die with the component during ejection.

Ejector die   The half of a casting die set that is mounted to the toggle and clamp side of the die casting machine.

Ejector pins   Movable pins in the die that help eject the component once solidification has occurred.

Erosion   The wearing of the die surface due to repetitive molten metal flow.

Flash A thin section of metal that forms at gaps along the parting line or between other die components.

Flux A metal-refining chemical used to control chemistry and remove undesired substances prior to casting.

Gas porosity A condition existing in a component caused by the trapping of gas in the molten metal during the manufacture of the component.

Gate The portion of the runner where the molten metal enters the die cavity.

Heat A single charge of metal.

Heat checking veins A surface defect that is caused by molten metal filling small cracks on the face of the die cavities caused by repetitive thermal cycling.

Heat treatment A combination of thermal treatments timed and applied to an alloy in the solid state in a specific manner producing desired mechanical properties.

Hot-chamber die casting The die casting process variant in which the metal injection system is continuously immersed in the molten metal.

Inclusions Particles of slag, refractory materials, or oxidation products trapped in a component during pouring and solidification.

Indirect semi-solid metalworking All semi-solid metalworking process variants in which solid feedstock with a spheroidal microstructure is manufactured and then reheated for use in manufacturing products.

Induction heating A method of heating in which an alternating current and coil are used to create a secondary current within the metal causing heat to be generated.

Ladle A cup used to transfer molten metal from the holding furnace to the shot sleeve.

Laminar flow Flow of unidirectional movement with no mixing occurring within the fluid.

Locating pad A projection on a component that helps maintain alignment of the component for machining operations.

Locating surface A surface to be used as a basis for measurement in making secondary machining operations.

Magneto-hydrodynamic stirring The use of an electromagnetic field to stir a molten metallic fluid.

Nonplanar flow Flow behavior prone to air entrapment in which the fill front is not uniform and progresses randomly in multiple directions.

Overflows Masses of metal cast with a product to aid in the control of filling, porosity, and other potential defects.

Parting line The line of separation for the independent die components.

Phase separation A nonhomogeneous microstructure that occurs in semi-solid metal, which forms when solid-phase particles and the liquid phase do not uniformly fill the die cavity.

Planar flow Flow behavior in which the fluid front progresses as a uniform surface in a consistent direction.

Returns Metal (of known composition) in the form of gates, runners, overflows, and scrapped component returned to the furnace for remelting.

Runner system The set of channels in a die through which molten metal flows to fill the die cavities.

Semi-solid billet/feedstock Material with a spheroidal microstructure that is reheated for use in indirect semi-solid metalworking processes.

Semi-solid metalworking A variant of the die casting process in which a partially liquid–partially solid metal mixture is injected into the die cavity.

Short shot Incomplete filling of the die cavity caused by an undersized volume of metal being metered into the metal injection system.

Shrinkage Contraction of metal in the die during solidification.

Shrinkage porosity A defect resulting from the volume difference that occurs when liquid metal solidifies.

Sinks A surface defect caused by localized solidification shrinkage beneath the surface of a component.

Sludge Intermetallic compounds that can precipitate in liquid alloys, causing hard spots in solidified components and altering the chemical composition of the alloy.

Slurry A flowable mixture of particles suspended in a liquid.

Soldering The bonding of a component during solidification to the die surface often resulting in permanent damage to both the component and the die.

Spheroidal structure The spherically shaped metal microstructure observed in products manufactured using semi-solid metalworking processes.

Squeeze casting A high integrity die casting process characterized by the use of large gate areas and planar filling of the metal front within the die cavity.

Trimming Removal of runners, risers, flash, overflows, and other surplus metal from a component.

Turbulent flow Flow behavior in which macroscopic mixing occurs within the fluid.

Vacuum die casting A die casting process that utilizes a controlled vacuum to extract gases from the die cavities and runner system.

Vent An opening or passage in a die or around a core to facilitate the escape of gases.

APPENDIX B
RECOMMENDED READING

B.1 BOOKS

The following books are recommended by the author as reference in the areas of die casting processing, die casting design, casting defects, and metal solidification. Although some date back over a decade, the information disclosed in these books provides basic information about casting processes in general as well as historical information regarding the evolution of die cast processes.

1. *Analysis of Casting Defects*, 4th ed., American Foundry Society, Des Plaines, IL, 1974.
2. *Magnesium Die Casting Handbook*, North American Die Casting Association, Rosemont, IL, 1998.
3. Stefanescu, D. (Ed.), *Metals Handbook,* 9th ed., Vol. 15: *Casting,* ASM International, Materials Park, OH, 1988, p. 327.
4. *NADCA Product Specification Standards for Die Castings,* North American Die Casting Association, Rosemont, IL, 1997.
5. *NADCA Product Specification Standards for Die Castings Produced by the Semi-Solid and Squeeze Cast Processes,* North American Die Casting Association, Rosemont, IL, 2000.
6. *Product Design for Die Casting,* 5th ed., Die Casting Development Council of the North American Die Casting Association, Rosemont, IL, 1998.

7. Doehler, H., *Die Casting,* McGraw-Hill, New York, 1951.
8. Flemings, M., *Solidification Processing,* McGraw-Hill, New York, 1974.
9. Jorstad, J., and W. Rasmussen, *Aluminum Casting Technologies,* 2nd ed., American Foundry Society, Des Plaines, IL, 1993.
10. Kaye, A., and A. Street, *Die Casting Metallurgy,* Butterworth Scientific, London, 1982.

B.2 PAPERS

Currently, a conference dedicated to semi-solid metalworking technologies is held on an annual basis. The proceedings from these conferences are excellent resources for keeping pace with this evolving technology. In addition to these proceedings, the following papers are recommended for review by the author.

1. Alexandrou, A., G. Burgos, and V. Entov, "Semisolid Metal Processing: A New Paradigm in Automotive Part Design," SAE Paper Number 2000-01-0676, Society of Automotive Engineers, Warrendale, PA, 2000, p. 15.
2. Flemings, M., "Behavior of Metal Alloys in the Semisolid State," *Metallurgical Transactions,* Vol. 22B, June 1991, p. 269.
3. Merens, N., "New Players, New Technologies Broaden Scope of Activity for Squeeze Casting, SSM Advances," *Die Casting Engineer,* 1999, p. 16.
4. Reynolds, O., "An Experimental Investigation of the Circumstances Which Determine Whether Motion of Water Shall Be Direct or Sinuous and of the Law of Resistance in Parallel Channels," *Transactions of the Royal Society of London,* Vol. A174, the Royal Society of London, London, England, 1883, p. 935.
5. Wolfe, R., and R. Bailey, "High Integrity Structural Aluminum Casting Process Selection," SAE Paper Number 2000-01-0760, Society of Automotive Engineers, Warrendale, PA, 2000.

6. Young, K., "Semi-Solid Metal Cast Automotive Components: New Markets for Die Casting," Paper Cleveland T93-131, North American Die Casting Association, Rosemont, IL, 1993.

B.3 PERIODICALS

Review of the following metals industry publications is recommended to keep pace with the latest developments in semi-solid metalworking technology.

1. *Advanced Materials and Processes,* ASM International, Materials Park, OH.
2. *Die Casting Engineer,* North American Die Casting Association, Rosemont, IL.
3. *Engineered Casting Solutions,* American Foundry Society, Des Plaines, IL.
4. *Light Metal Age,* Fellom Publishing, South San Francisco, CA.
5. *Modern Casting,* American Foundry Society, Des Plaines, IL.
6. *Modern Metals,* Trend Publishing, Chicago, IL.

APPENDIX C
MATERIAL PROPERTIES OF ALUMINUM

A committee of The North American Die Casting Association has worked to compile material property information using standard samples for conventional die casting (CDC), semi-solid metalworking (SSM), and squeeze casting.[1,2] Material properties, however, vary in actual components due to disparities in geometry, microstructure, and thermal history. The typical material data collected using standard samples should be used only as a guide. The data presented in this appendix was collected by measuring the material properties of production components.

REFERENCES

1. *NADCA Product Specification Standards for Die Castings,* North American Die Casting Association, Rosemont, IL, 1997.
2. *NADCA Product Specification Standards for Die Castings Produced by the Semi-Solid and Squeeze Cast Processes,* North American Die Casting Association, Rosemont, IL, 2000.
3. DasGupta, R., and D. Killingsworth, "Automotive Applications Using Advanced Aluminum Die Casting Processes," SAE Paper Number 2000-01-0678, Society of Automotive Engineers, Warrendale, PA, 2000.

TABLE C.1 Tensile Properties of Common Die Cast Aluminum Alloys

Alloy	Process	Yield Strength (MPa)	Tensile Strength (MPa)	Elongation (%)
280-T5	CDC	NA	152–165	1.2–1.4
280-F	Squeeze	NA	214–234	2–3
A356.2-T6	Squeeze	145–165	255–276	13–17
A356.2-T6	SSM	152–168	261–284	16–20
357-T6	Squeeze	241–262	324–338	8–10
357-T6	SSM	237–257	315–330	7–9
390-F	CDC	241	279	<1
390-T6	Squeeze	NA	352–392	<1
390-T6	SSM	NA	341–386	<1
319-T4[a]	SSM	182	308	7.3
319-T6[a]	SSM	293	361	3.7

Note: All tensile specimens machined from actual castings. NA = not available.
[a] EM-stirred aluminum alloy.
Source: From Ref. 3. Courtesy of SPX Contech Corporation.

TABLE C.2 Impact Properties of Common Die Cast Aluminum Alloys

Alloy	Process	Temper	Elongation (%)	Impact Strength (J)
380	CDC	T5	1.2–1.4	<1
380	Squeeze	F	2–3	1–3
A356.2	Squeeze	T6	13–17	14–18
A356.2	SSM	T6	16–20	17–20
357	Squeeze	T6	8–10	10–13
357	SSM	T6	6–9	9–11
390	CDC	F	<1	<1
290	Squeeze	T6	<1	<1
390	SSM	T6	<1	<1
319[a]	SSM	T4	7.3	26
319[a]	SSM	T6	3.7	19

Note: Cross section of impact specimens is 10 × 3.3 mm; unnotched; all specimens machined from actual castings.
[a] EM-stirred aluminum alloy.
Source: From Ref. 3. (Courtesy of SPX Contech Corporation.)

TABLE C.3 Wear and Cavitation Resistance of Common Die Cast Aluminum Alloys

Alloy	Wear,[a] $\times 10^{-12}$ (m³)	Cavitation Erosion,[b] $\times 10^{-10}$ (m³)
CDC 380-F	462–16,000	NA
A356.2-T6[c]	173–434	NA
A356.2-T6[d]	180–455	13.5 (machined)
		21.8 (as cast)
357-T6[d]	166–347	NA
357-T6[c]	140–363	NA
CDC 390-F	125–234	NA
390-T6[d]	55–97	NA
390-T6[c]	50–100	NA

[a] Volume loss of material; test performed according to ASTM G77.
[b] Volume loss of material; testing consisted of a single 30-min exposure of a 35 × 35 × 1.5-mm sample to the tip of an ultrasonic horn operating at 20 kHz; the vibrating horn was located 0.5 mm above the sample submerged in a beaker of water. NA = not available.
[c] Squeeze casting.
[d] Semi-solid metalworking.
Source: From Ref. 3. Courtesy of SPX Contech Corporation.

TABLE C.4 Impact Resistance and Properties of Common Die Cast Aluminum Alloys

Alloy	Process	Temper	Elongation %	Impact Strength (J)
380	CDC	T5	1.2–1.4	<1
380	Squeeze	F	2–3	1–3
A356.2	Squeeze	T6	13–17	14–18
A356.2	SSM	T6	16–20	17–20
357	Squeeze	T6	8–10	10–13
357	SSM	T6	7–9	9–11
390	CDC	F	<1	<1
390	Squeeze	T6	<1	<1
390	SSM	T6	<1	<1
319[a]	SSM	T4	7.3	26
319[a]	SSM	T6	3.7	19

Note: Cross section of impact specimens is 10 × 3.3 mm; unnotched; all specimens machined from actual castings.
[a] EM-stirred aluminum alloy.
Source: From Ref. 3. Courtesy of SPX Contech Corporation.

TABLE C.5 Fracture Toughness of Common Die Cast Aluminum Alloys

Alloy	Process	Temper	Elongation (%)	K_Q [MPa$(m)^{1/2}$]
A356.2	Squeeze	T6	13–17	18.7–22.5
A356.2	SSM	T6	16–20	19.5–23.5
319[a]	SSM	T4	7.3	22.8
319[a]	SSM	T6	3.7	22.9

Note: Samples machined from actual castings. All samples, followng machining, were precracked and tested at 24°C per ASTM E399-90 and ASTM B645.
[a] EM-stirred aluminum alloy.
Source: From Ref. 3. Courtesy of SPX Contech Corporation.

TABLE C.6 Fatigue Properties of Common Die Cast Aluminum Alloys

Alloy	Process	Temper	Fatigue Strength (MPa)
A356.2	Squeeze	T6	106[a]
A356.2	SSM	T6	117[b]
319	SSM	T4	164[c]

[a] Mean fatigue strength based on "staircase" method at 10^7 cycles; axial loading; $R = -1.0$.
[b] Fatigue strength at 10^7 cycles from S–N curve; axial loading; $R = -1.0$.
[a] Mean fatigue strength based on staircase method at 10^7 cycles; axial loading; $R = -1.0$.
Source: From Ref. 3. Courtesy of SPX Contech Corporation.

APPENDIX D
DIE CAST MAGNESIUM MATERIAL PROPERTIES

Magnesium (Mg) has a density of 1.74 g/cm^3 giving it a significant weight advantage over aluminum. The typical material data presented in this appendix have been developed using "as-cast" specimens. Material properties will vary in actual components due to variations in geometry, microstructure, and thermal history.

TABLE D.1 Material Properties for Magnesium Alloys

	Magnesium Die Casting Alloys						
	AZ91D	AZ81	AM60B	AM50A	AM20	AE42	AS41B
			Mechanical				
Ultimate tensile strength[a]							
ksi	34	32	32	32	27	33	31
MPa	230	220	220	220	185	225	215
Yield strength[a,b]							
ksi	23	21	19	18	15	20	20
MPa	160	150	130	120	105	140	140
Compressive yield strength[c]							
ksi	24	n/a	19	n/a	n/a	n/a	20
MPa	165		130				140
Elongation[a]							
% in 2 in. (51 mm)	3	3	6–8	6–10	8–12	8–10	6
Hardness,[d]							
BHN	75	72	62	57	47	57	75
Shear strength[a]							
ksi	20	20	n/a	n/a	n/a	n/a	n/a
MPa	140	140					
Impact strength[e]							
ft-lb	1.6	n/a	4.5	7.0	n/a	4.3	3.0
J	2.2		6.1	9.5		5.8	4.1
Fatigue strength[f]							
ksi	10	10	10	10	10	n/a	n/a
MPa	70	70	70	70	70		
Latent heat of fusion							
Btu/lb	160	160	160	160	160	160	160
kJ/kg	373	373	373	373	373	373	373
Young's modulus[a]							
psi × 10^6	6.5	6.5	6.5	6.5	6.5	6.5	6.5
GPa	45	45	45	45	45	45	45

Physical

Density[a]							
lb/in.³	0.066	0.065	0.065	0.064	0.063	0.064	0.064
g/cm³	1.81	1.80	1.79	1.78	1.76	1.79	1.77
Melting range							
°F	875–1105	915–1130	1005–1140	1010–1150	1145–1190	1050–1150	1050–1150
°C	470–595	490–610	540–615	543–620	618–643	585–620	565–620
Specific heat[a]							
Btu/lb°F	0.25	0.25	0.25	0.25	0.24	0.24[a]	0.24
J/kg°C	1050	1050	1050	1050	1000	1000	1020
Coefficient of thermal expansion[a]							
μin./in./°F	13.8	13.8	14.2	14.4	14.4	14.5[a]	14.5
μm/m K	25.0	25.0	25.6	26.0	26.0	26.1	26.1
Thermal conductivity							
Btu/ft hr °F	41.8[h]	30[a]	36[a]	36[a]	35[a]	40[a,g]	40[a]
W/m K	72	51	62	62	60	68	68
Electrical resistivity[a]							
μΩ in.	35.8	33.0	31.8	31.8	n/a	n/a	n/a
μΩ cm	14.1	13.0	12.5	12.5			
Poisson's ratio	0.35	0.35	0.35	0.35	0.35	0.35	0.35

Note: Casting conditions may significantly affect mold shrinkages. n/a = data not available.

[a] At 68°F (20°C).
[b] Offset 0.2%.
[c] Offset 0.1%.
[d] Average hardness based on scattered data.
[e] ASTM E 23 unnotched 0.25 in. die cast bar.
[f] Rotating beam fatigue test according to DIN 50113. Stress corresponding to a lifetime of 5×10^7 cycles.
[g] Estimated.
[h] At 212–572°F (100–300°C).

Source: From Ref. 1. Courtesy of the North American Die Casting Association.

REFERENCE

1. *NADCA Product Specification Standards for Die Castings,* North American Die Casting Association, Rosemont, IL, 1997.

INDEX

Actuated shut-off valves, *see* Vacuum shut-off valves
Aluminum alloy freezing ranges, 67–68
Aluminum alloys, 5, 22–23, 30, 45–46, 52, 60–63, 67, 71, 82, 84, 86–91, 102, 115–116, 159, 171–175
Atomized flow, *see* Liquid metal flow

Bell laboratories, 147
Blistering, 10, 32, 34, 42–43, 47, 57, 61, 82, 162
Brass alloys, 5

Casting dies
  gating, 11, 19, 21, 51, 53–56, 70, 75–76, 117, 119, 141, 164–167
  runner systems, 6, 11, 29–30, 36–37, 56, 58, 74, 76–77, 80, 104, 107, 141, 159, 162, 165, 167
Centerline porosity, 160
Chill-block shut-off valve, *see* Vacuum shut-off valves
Chilled vents, *see* Vacuum shut-off valves
Clausing, Don, 132

Cold shuts, 158
Common-cause variation, 146–149, 153
Component integration, *see* Integration
Composites, *see* Metal matrix composites
Computer modeling, *see* Simulations
Contaminant veins, 24, 32, 105, 163–165
Contamination, 160
Control factor flexibility curve, 132
Controlled porosity, 177–178
Corrugated chill block, *see* Vacuum shut-off valves
Cracks, 8, 102, 158, 162

Defects
  dimensional, 161
  internal, 159–161
  prediction of, 167–168
  from secondary processing, 161–162
  surface, 158–159
  unique to semi-solid metalworking, 162–167
  unique to squeeze casting, 162–165
Design for assembly, 127

219

Design for manufacturability
  case study, 114–122
  defined, 113
  rules for high integrity die casting processes, 114
Die casting
  casting cycle, 6, 8
  cold chamber, 5, 7
  conventional high pressure, 3–5
  cycle times, 5, 30, 34, 54, 57, 70, 82, 105, 107
  hot chamber, 5–6
  limitations of, 7–10
  origins, 3–5
  problems with, 7–10
  strategies for improvement, 10
Die filling, *see* Metal flow
Die lubricant, 6, 9, 29–30, 53, 70, 101, 104–106, 152, 158–161, 163, 165
Die surface temperature, 102, 151–155
Die thermal management, 101, 105–108
Dies, *see* Casting dies
Discoloration, 158
Doehler, H.H, 3–5
Drags, 158

Electric shut-off valves, *see* Vacuum shut-off valves, dynamic
Emerging alloys, 173
Erosion, 161

Fins, *see* Heat checking
Flash, 158
Flux contamination, 160

Gas porosity, 29, 34, 53–54, 70, 160
Gutenberg, Johannes, 3

Hard spots, 161
Hawthorne works, 145
Heat checking, 102–106, 159

Heat treating, 10–11, 32, 34, 42–43, 47, 51, 57–62, 82, 88–89, 91, 162
  aluminum T5, 91
  aluminum T6, 43, 47, 60, 61, 89
Hydraulic shut-off valves, *see* Vacuum shut-off valves, dynamic

Inclusions, 160
Individuals chart, 150
Inherent variation, *see* Common-cause variation
Integration, 86, 125–130
  analysis flow chart, 128
  case study, 129–130
  costs, 125–127
Interdendritic porosity, 160
Intermittent variation, *see* Special-cause variation

Laminar flow, *see* Liquid metal flow, laminar
Laminations, 159
Lead alloys, 5
Life cycle cost lever, 133
Linotype machine, 3
Liquid metal flow
  atomized, 15, 19–20, 32, 53
  in conventional die casting, 19–20
  at the fill front, 15–19
  within a fluid, 13–15
  laminar, 13–15
  nonplanar, 16–19
  planar, 16, 17
  predicting, 24
  in semi-solid metalworking, 22–23
  in squeeze casting, 21–22
  turbulent, 13–15
  in vacuum die casting, 19–21
Lost-core technology, 176–177

Magnesium alloy development, 172–173

INDEX **221**

Magnesium alloy microstructures, 80, 82
Mechanical shut-off valves, *see* Vacuum shut-off valves, dynamic
Mergenthaler, Ottmar, 3
Metal Flow, *see* Liquid metal flow
Metal matrix composites, 173–175
Microstructure of die welds, 103, 104

Nonplanar flow, *see* Liquid metal flow, nonplanar

Phase separation, 165–167
Planar filling, *see* Liquid metal flow, planar
Plunger lubricant, 6
Porosity, 160
  gas, 29, 34, 53–54, 70, 160
  quantified, 8–9
  shrinkage, 7–9, 11, 35, 42, 61, 70, 117–121, 135
  sources of gas porosity, 9
Powder die lubricants, 106–108
Powder lubricant application, 107–108
Process data curves, 150–152
Process simulations, *see* Simulations
Product integration, *see* Integration

Reducing tool lead times, 175–176
Reynolds number, 14–15
Reynolds, Osborne, 14
Rheocasting, *see* Semi-solid metalworking
Rotary vane vacuum pumps, 35–36

Semi-liquid casting, *see* Semi-solid metalworking
Semi-liquid metalworking, *see* Semi-solid metalworking

Semi-solid billet
  anatomy, 77–78
  behavior during heating, 77–78
  dendritic case, 77
  extruded, 77–79
  magneto-hydrodynamic stirring, 76
Semi-solid casting, *see* Semi-solid metalworking
Semi-solid metalworking
  aluminum, 67–68, 71–74
  case studies, 84–100
  comparison to conventional die casting, 82–83
  compatible metal systems, 67
  die design, 75, 80–81
  direct process, 79
  direct process manufacturing cell layout, 79–80
  equipment, 72–82
  heat treatability, 82–83
  indirect process, 73–79
  indirect process manufacturing cell layout, 73–75
  magnesium, 70–82
  microstructure (typical), 72–74
  origins, 67
  proper application of, 82–83
  spheroidal microstructure, 71–74
Shewhart, Walter, 145–147
Shrinkage porosity, *see* Porosity
Short shot, 159
Simulations
  case studies, 134–138, 140, 141
  commitment, 140
  common types, 134
  effectiveness, 138–139
  of high integrity die casting processes, 134–140
  planning, 138–139
  resources required, 139–140
Sinks, 159
Sludge, 160–161
Soldering, 102–105, 159
Solidification, 3

Solidification shrinkage porosity, *see* Porosity
Special-cause variation, 146–153
Squeeze casting
  case studies, 58–64
  comparison to conventional die casting, 57–58
  equipment, 56–57
  heat treatability, 57–58
  microstructure (typical), 53–54
  proper application of, 57–58
Squeeze forming, 51–52
Statistical process control
  applied to high integrity die casting processes, 154–155
  case study, 151–154
  dartboard comparison of variation, 146–147
  of dynamic process characteristics, 149–151
  identifying special-cause variation, 149–150
  origins, 145–148
  out-of-control signals, 149–150
  process, 148
  product, 148
Strategies for containing porosity, 10–11
Surface staining, 158
Surface veins, *see* Heat checking

Thermal balancing of dies, 105–106, 108
Thermal cycling of dies, 101–102
Thixocasting, *see* Semi-solid metalworking
Thixomolding®
  case studies, 91–100
  machine, 79
  magnesium microstructures, 80, 82
  manufacturing cell layout, 79–80
  process, 79
  runner design, 80–81
Titanium alloys, 173
Turbulent flow, *see* Liquid metal flow

Vacuum die casting
  case studies, 42–49
  equipment, 35–40
  proper application of, 40–42
Vacuum pump operating curve, 35
Vacuum pumps, 35–37
Vacuum shut-off valves
  casting cycle, 32–34
  dynamic, 38–40
  effectiveness, 38–39, 41
  maintenance, 38
  placement, 32–33
  static, 37–38
  timing, 32–34
Vacuum valves, *see* Vacuum shut-off valves
Valves, *see* Vacuum shut-off valves
Venting
  effectiveness, 29–32
  strategies, 32
Viscosity
  of semi-solid aluminum, 23
  of liquid aluminum, 20, 22

Warpage, 161
Welding of dies, 103–104
Western Electric Company, 145, 147

X chart, 150
x-bar chart, 149

Zinc alloys, 3, 57, 62, 67

ABOUT THE AUTHOR

Edward J. Vinarcik, P.E., is a product engineer working for a major tier 1 automotive supplier. In 1993, he received his bachelor's degree in metallurgical engineering from The Ohio State University, where he conducted die cast research as an undergraduate fellow. He continued his studies earning a masters of science in quality and an MBA. He has studied under Drs. Akao and Taguchi. He is certified by the American Society for Quality as a Quality Manager and Quality Engineer. Prior to working in the automotive supply base, he served as both a die cast manufacturing engineer and product design engineer with Ford Motor Company. While serving in these capacities, Mr. Vinarcik's work focused on the design, development, and manufacture of automotive fuel components using squeeze casting and semi-solid metalworking processes. In addition to writing technical papers and articles, he serves on numerous committees, including the Editorial Review Board for *Advanced Materials and Processes* magazine, the Ground Transportation Industry Sector Committee of ASM International, and the Society of Automotive Engineers' Materials Engineering Activity. His efforts have been recognized by the Engineering Society, which presented him the "Outstanding Young Engineer Award" in 2002.

CUSTOMER NOTE: IF THIS BOOK IS ACCOMPANIED BY SOFTWARE, PLEASE READ THE FOLLOWING BEFORE OPENING THE PACKAGE.

By opening the package, you are agreeing to be bound by the following agreement:

This software product is protected by copyright and all rights are reserved by the author, John Wiley & Sons, Inc., or their licensors. You are licensed to use this software as described in the software and the accompanying book. Copying the software for any other purpose may be a violation of the U.S. Copyright Law.

This software product is sold as is without warranty of any kind, either express or implied, including but not limited to the implied warranty of merchantability and fitness for a particular purpose. Neither Wiley nor its dealers or distributors assumes any liability for any alleged or actual damages arising from the use of or the inability to use this software. (Some states do not allow the exclusion of implied warranties, so the exclusion may not apply to you.)

© 2003 John Wiley & Sons, Inc.